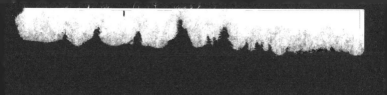

D1485704

On the
EDGE
of
INFINITY

An Hachette UK Company
www.hachette.co.uk

First published in Great Britain in 2018 by Cassell, an imprint of
Octopus Publishing Group Ltd
Carmelite House
50 Victoria Embankment
London EC4Y 0DZ
www.octopusbooks.co.uk

Originally published in Germany as:
"Das All und Das Nichts. Von der Schönheit des Universums"
© S. Fischer Verlag GmbH, Frankfurt am Main, 2017

Copyright © Octopus Publishing Group 2018

All rights reserved. No part of this work may be reproduced or utilized in any form
or by any means, electronic or mechanical, including photocopying, recording or by
any information storage and retrieval system, without the prior written permission
of the publisher.

Stefan Klein asserts the moral right to be identified as the author of this work.

ISBN 978 1 78840 0 602

A CIP catalogue record for this book is available from the British Library.

Printed and bound in Great Britain

10 9 8 7 6 5 4 3 2 1

Translated by Mike Mitchell

Publishing Director: Trevor Davies
Senior Editor: Alex Stetter
Junior Designer: Jack Storey
Proofreader: Guy Tindale
Cover design: Schiller Design, Frankfurt, adapted by Jack Storey
Typesetter: Jeremy Tilston
Senior Production Manager: Peter Hunt

The translation of this work was supported by a grant from the Goethe-Institut.

On the
EDGE
of
INFINITY

Uncovering the Visible World's Scientific Beauty

STEFAN KLEIN
Germany's bestselling science author

DEDICATION

In memory of my father,
who showed me

CONTENTS

1 – The Poetry of Reality   11

*A rose makes us aware
that nothing and nobody stands alone.
The more we know about how things in the universe relate
to each other, the more mysterious the world seems to us.*

2 – A Marble in the Cosmos   21

*The Earth rises over the Moon and we see
the universe as it is being born. Much greater
spaces are concealed behind the visible cosmos.
Reality is quite different from how it seems to us.*

3 – Riding on a Ray of Light   37

*A young man wonders what light is,
and his reflections on light explain the world
to him. Time and space are revealed.
But when Albert Einstein dies,
light is still a mystery.*

4 – The World Spirit Fails   57

*A hurricane sweeps across Germany,
a storm no one saw coming.
Reasons why the world is unpredictable,
and praise of the creative universe.*

5 – A Crime Story 79

*A villainous gang is raiding flats in
London and New York.
Although the burglars were not able to arrange things
with each other, their raids are perfectly coordinated.
Investigator Glock is looking for a secret plan,
but cannot find one. His conclusion:
all the places in the world are in reality one place.*

6 – Is the World Real? 109

*A hammer hits a thumb. But the hammer,
like all matter, consists of emptiness.
How can nothingness hurt like that?
And then – does the nothingness exist at all?*

7 – 'Who ordered that?' 127

*We live in a shadow world. No matter
where we look, there is twenty times more than
appears to us. More of what? We have no idea.
But without dark energy, without dark matter
we couldn't exist.*

8 – How Time Passes 147

*A greying beard makes you wonder why the
past can never come back. We experience
the passing of time because we are not omniscient.
The universe is growing older as well.*

9 – Beyond the Horizon   167

The night is dark
because the world had a beginning.
Since then the universe has been expanding.
Space is bigger than we can imagine.
Thoughts on being amazed.

10 – Why we exist   187

In each of us one of the most astonishing
characteristics of the universe appears:
intelligent life is not only possible but even probable.
How can anyone maintain therefore
that we are meaningless?

Notes   211

Thanks   237

About the author   239

I

The Poetry of Reality

*A rose makes us aware
that nothing and nobody stands alone.
The more we know about how things in the universe relate
to each other, the more mysterious the world seems to us.*

The more we know about reality, the more mysterious it seems to us. Astonishingly enough, it is sensitive people in particular who dispute that. During a panel discussion, a well-known German poet once remonstrated with me, saying that he detested our ever more precise knowledge of genes because decoded man was a bore. And Edgar Allan Poe, the master of the mystery story, called science a predator on poetry:

> *Why preyest thou thus on the poet's heart,*
> *Vulture, whose wings are dull realities?*

How wrong one can be! Poets are rightly afraid of a world that has lost its magic, but anyone who harbours that fear is confusing research into our world with an Easter egg hunt, in the course of which all the hiding places are eventually plundered. Genuine insight, however, regularly throws up more questions than it can answer.

The great American physicist Richard Feynman was once asked by an artist friend whether a scientist wouldn't destroy the beauty of a rose if he examined it. Feynman replied that he was able to feel the beauty the artist felt, certainly, but that he saw a deeper beauty, one that only revealed itself through understanding – for example in the fact that flowers acquired colour during the course of evolution in order to attract insects. This knowledge, he went on, led to further questions, for example whether insects had any kind of aesthetic sense. Getting to know the flower more closely took away none of its beauty – on the contrary, it added beauty, and made the rose appear even more impressive and mysterious.

Feynman could have gone on to say that the scientist's sharp eye even revealed beauty in things that at first seem ugly or even repulsive to us. The fading of the rose is a symbol of decline, but if you look closely, you can see the hip growing deep within the withering petals. Each seed in the rose hip is a miracle of its own, because in each tiny kernel, the complete embryo of a rose is waiting for the moment when it can soak up water, expand, break out of the husk and stretch out its seed-leaves to the sun.

In order to grow, the germinating rose needs light, water and oxygen. Living beings from long ago have bequeathed it air to breathe. The flower is heir to single-celled organisms that covered the sea-bed in thick, blue-green mats well over three billion years ago and still live there today. Back then there was almost no oxygen in the atmosphere, and all higher life-forms would have suffocated. The single-celled organisms were only a few thousandths of a millimetre in size. Compared with the rose, these creatures called cyanobacteria seem exceedingly primitive to us and yet they were already masterpieces of nature. Some cyanobacteria can even see! Their bodies have a tiny receptor, a simple camera eye that allows them to distinguish between light and darkness. They avoid the darkness and move towards the light. They use the sunlight in order to acquire energy through photosynthesis, like modern plants. After cyanobacteria had settled in the ancient ocean, they converted the carbon dioxide dissolved in the sea-water into oxygen. For a billion years, the oxygen bubbled up from the depths of the ocean. Thus these sighted cyanobacteria created the air that the rose needs to germinate. They made the Earth habitable for higher life forms.

For their part, the cyanobacteria developed out of earlier, even simpler life forms that could also survive without oxygen. These unknown organisms colonized the Earth 3.8 billion years ago. Without them we would never have had a chance to see a rose. Where did this life come from? That we don't know.

And where does the rose get its water from? That too has its own story and goes further back than the story of the air. For a long time, we were content with the observation that in the early period of our planet, steam had come out of the interior of the Earth as a gas. But how did the water get into the centre of the Earth in the first place? It could only have been locked up in there when the Earth was formed. 4.5 billion years ago, lumps of stone and dust that were revolving around the Sun combined to form the planets; the Earth was formed from material that was moving not far from the Sun. However, it is practically impossible that this debris was damp enough to make Earth into the blue planet – the heat of the nearby Sun would certainly have dried it up.

So originally the Earth was presumably dry, a desert planet. We don't know precisely how it turned into a world of oceans. The scenario that seems to be the most fantastic of all possible explanations is in fact the most likely one: water came to us from outer space. It arrived in comets or asteroids that, born in the colder regions of the solar system, hit the desert planet Earth like gigantic snowballs. Thus the lakes, rivers and oceans were filled with the melted ice from the comets. Dewdrops from the cosmos moisten the leaves of the rose.

Finally, the rose owes the strong force its light. The name of this elemental force is actually too modest, for the strong force is far and away the strongest in nature. It holds atomic nuclei together. It is released in the interior of the Sun, where the atomic nuclei of hydrogen fuse to become helium, releasing immense amounts of energy that radiate out into space. Hydrogen, that combustible material, is the oldest of all substances. Since the very first minute after the Big Bang, hydrogen has been present throughout the cosmos. All the elements were baked from it in the furnace of the stars, again as a result of the strong force. Everything around us on Earth was once the ash of the stars – the carbon of which the germ bud consists also comes from it. The rose is metamorphosed stardust.

The stars that brought forth the rose, however, were born of clouds of hydrogen. As a result of their own gravitational pull in the cosmos, those clouds became so concentrated that they eventually ignited – the first starlight shone out. Did the stars then give birth to themselves? For a long time that was the assumption. Today we know: the stars too needed outside help. The hydrogen in the universe wasn't sufficient to condense into clouds by means of its own gravitational pull. Left to itself, it would simply have spread evenly over the cosmos, like sugar dissolving in tea. The gases would never have concentrated, not even a single star would ever have shone in the sky. The universe would have remained without form.

Something heavy must therefore have started it all by attracting the hydrogen, causing it to form clouds – something we don't know. As this something doesn't shine and also remains otherwise invisible, we call it 'dark matter'. What this dark matter consists of, what characteristics it has, we do not know.

Many of these connections were unknown to Richard Feynman, who reflected on the beauty of the rose. He died in 1988, one of the most important scientists of the 20th century. But in recent years our knowledge of the formation of the world has expanded dramatically. We are now able to trace the history of the universe back to the first millionth of a second after its birth, at least in broad outline. We know of habitable planets outside the solar system, we have discovered a system 40 light years away with seven planets similar to Earth and must assume that the night sky conceals many more planets than the stars that shine. We are aware of physical processes that are contrary to our notions of space and time.

Even very recently, the idea that knowledge of this kind might be possible was still regarded as adventurous speculation. Today it consists of facts that are substantiated by measurements right down to the decimal point.

But our knowledge is merely an island in an ocean of ignorance. And whenever we manage to extend the island, we also lengthen the coastline from which we confront our lack of knowledge. Despite all the spectacular insights we have gained, the questions have not become any fewer, and certainly not simpler. We would love to know what took place in the first millionth of a second after the birth of the universe. And does it make sense to speculate on what might have been going on previously, before the Big Bang? Is there indeed life elsewhere in the cosmos? Are time and space merely illusions? These are the kinds of questions this book deals with. It describes how 21st-century physics changes our thinking, the way we see the world. It doesn't demand any prior knowledge, only the courage to look behind the veil of that which still seems self-evident to us today. Then a world will be revealed to us that 'is not only queerer than we suppose but queerer than we *can* suppose,' as the British biologist J B S Haldane put it. The following pages are an invitation to let yourself be enthralled by the reality in which we live. For a rose is much more than a rose. It is a witness to the genesis of the world.

2

A Marble in the Cosmos

*The Earth rises over the Moon and we see
the universe as it is being born. Much greater
spaces are concealed behind the visible cosmos.
Reality is quite different from how it seems to us.*

...for all knowledge and wonder is an impression
of pleasure in itself.
FRANCIS BACON

In one of my earliest childhood memories, my father is carrying a large cardboard box into the house. He pushes his way in through the door backwards, then the box appears, finally a friend of my father's, who has grabbed hold of the other end. 'What is it?' my mother asks. 'I've bought us a television,' he replies. My mother is furious, she doesn't want such an ugly applicance in the house. My father justifies himself: 'They're flying to the Moon.'

My father fetches a saw. He opens the double doors of the dark cupboard in the living room, which we children aren't allowed to touch because it's a valuable old piece of furniture. Moreover, it's used as the liquor cabinet. My father takes out the bottles and gets to work inside the cupboard. He cuts out shelves from the precious cupboard until there's enough room for the television. When the doors are shut, it disappears.

So the astronauts' voices came out of the

cupboard. What sticks in my mind is the metallic tone in which they rasped out their commands, which were incomprehensible for me. I also remember two images. In one scene, two figures are scurrying across the screen. They're shining ghostly white and instead of faces, they have a disc at the front of their heads. There's a flag hanging in the greyness behind them. The ghosts are carrying huge backpacks but they're skipping and jumping as if they weighed nothing at all. My parents say something about gravity being six times less on the moon; I'd love to try that myself. I'm four years old.

The other scene shows a marble precisely in the middle of the picture, half illuminated, hovering in complete darkness. Even though we had a black-and-white TV at the time, I remember it being such a deep blue that the intensity of the colour was almost painful. Clearly photos I saw later on in magazines and books have superimposed themselves on the TV pictures in my memory and added colour to them. There are swirls of white shimmering over the blue and on the left-hand side a large, sharply outlined patch of brown can be seen. However, the whole foreground of the image is filled with a desert of monotonous ochre. Hills and craters stretch as far as the horizon of a range of mountains, above which is the marble. It is impossible to imagine that anything has ever lived, or will ever live in that ochre wasteland.

Thus Apollo 11 transmitted the Earth rising over the Moon to our two-hundred-year-old living-room cupboard. I can't remember how I reacted when these images flickered across our screen in July 1969, but every time I've seen them since, my feelings have grown stronger. So that is our home in the cosmos – a tiny ball, alone in unmeasurable night, fragile and beautiful. If you look very closely, you can even see the atmosphere, a filmy halo shimmering in the sunlight: the only home of life we know of, the only place where we can be.

For all that, there is no trace of human activity

to be seen on the blue sphere, nothing reminiscent of things that are familiar to us. The view from the Moon shows our habitat in a way we never normally consider it – from outside. And yet we immediately feel that this has to do with us. It is precisely this unfamiliar vantage point that gives the 'Earthrise' pictures their power. Once you've seen them, you can no longer take your own existence for granted. While we feel wrapped up in the monotony of our daily routine, life can certainly seem banal. But can there be anything more astonishing than this life when we realize that far and wide there is no sign of company, that we are lonely passengers on a speck of dust in the cold of the universe? For a deeper appreciation of this insight, we have to abandon our accustomed perspective.

People have repeatedly found that reality is quite different from the way it appears to us. The Earth is not flat, nor does the Sun revolve round it. The Moon is no light in the sky, but a mirror reflecting the rays of the Sun. The clouds we can see among the stars through a telescope aren't mist, but galaxies like our own. Animals and people didn't appear on the planet in their current form, they developed in the long course of evolution. Each of these insights was at one time considered outrageous. They contradicted everything people were able or willing to imagine.

Today we take these outrages for granted. The way we see the world today is based on them.

I have seldom seen this search for a new, more all-embracing view of reality put in a more memorable way than in a mysterious woodcut that appeared in 1888 in a book by the French astronomer Camille Flammarion. The picture, the origin of which is unknown, is generally called 'Traveller on the Edge of the World'. It shows a traveller leaving his familiar surroundings behind him to look in amazement on a cosmos of strange beauty. Behind the man we can see the world we are familiar with: gently rising hills with shrubs and trees, a village nestled by a lake on which the setting sun is shining. In the foreground it is already evening, the stars are twinkling in the sky. And it is this starry sky, which comes down to meet the earth, that the traveller has broken through with his upper body. His head is already in another world, a world beyond known phenomena. In it is the sparkle of fantastic swirls, clouds, wheels of fire, rays, lights; the man is stretching out his hand towards the wonders he encounters. But has the traveller really left the world that is familiar to him? A richly decorated frame holds the two parts of the picture together, and it is perhaps not mere chance that the swirls behind the firmament recall the lines of electromagnetic fields. Physicists had discovered these invisible lines of force two decades before the 'Traveller on the Edge of the World' was published.

Eyes wide with amazement, the traveller looks on another dimension of our existence. He is observing the phenomena behind familiar appearances – not those of an alien world, however, but of our own everyday one. That connects him with us when we watch the blue Earth rise over the Moon.

If I had to choose just one picture of the great discoveries of the 21st century, I would also take one from outer space. Admittedly, the star chart that the European space probe Planck sent us, which was published in 2013, is rather more difficult to read than the images of the traveller and the blue marble. At first sight – and in this it's similar to the swirling lines of the traveller on the edge of the world – it reminds us of an abstract painting. You can make out colourful dots, which combine to form a coherent pattern: continents perhaps?

Pictured are the entire heavens as visible from Earth. That it takes time for us to find our bearings in this display is no surprise, since it shows a picture of the world that is still unknown to most people. What we are seeing is a birth: the birth of the universe. The colours represent an emission released more than 13 billion years ago, shortly after the Big Bang – the first light of the world. It comes from everywhere, filling the whole cosmos. Over time the light has changed into heat radiation. Each colour in the star chart therefore represents a temperature: the afterglow of the Big Bang, which prevents space from ever going completely cold. In the empty void between the galaxies, there is still a last remnant of warmth. Even if our eyes cannot discern the background radiation, it can be received by a normal TV satellite dish.

Cosmic background radiation was discovered by chance in 1964, when physicists Arno Penzias and Robert Wilson were carrying out experiments in New Jersey with one of the first satellite antennas. They noticed interference that was coming almost uniformly from all directions. That meant it couldn't be coming from nearby New York City, and in fact wasn't of terrestrial origin at all. But its source couldn't be detected in the Milky Way, or in any other galaxy either. Penzias, who was born in Munich of Jewish parents and at the age of six had escaped the Nazis with his brother in one of the Kindertransports, explained the interference as coming from a 'white dielectrical substance' on the antenna – in plain English: bird shit. A pair of pigeons had made their nest in the antenna dish. Penzias and Wilson procured a pigeon trap, caught the birds, loaded them in a truck and released them 50km (30 miles) away from the antenna. The two pigeons came back. Eventually Penzias could see no other way out than to shoot the birds in the name of science. The antenna was cleaned, but the signal from nowhere remained. The physicists had no idea what was going on.

Then Penzias remembered the Big Bang theory. At the time, the idea that the universe originated in a 'big bang' and that the bang might even have left behind some radiation was regarded as wild speculation. To take this 'science fiction' seriously meant putting your whole career on the line. The cosmos was assumed to be eternal. But the radiation predicted by the Big Bang theory corresponded precisely to the signal from the antenna. The supposed interference from pigeon shit therefore turned out to be triumphant proof of the idea that the universe had a beginning; Penzias and Wilson were awarded the Nobel Prize for Physics in 1978.

Today the receiver with which the two physicists captured the signal from the Big Bang can be seen in the Deutsches Museum in Munich. Penzias presented it to the city of his birth, which he had had to leave as a child in order to save his life, in recognition of the fact that Germany hadn't denied its past but thrown light on it. 'I want to be part of this community, which is sharing its knowledge of the past with future generations,' he explained.

Fifty years after it was discovered by Penzias and Wilson, the first light of the cosmos was once more surveyed by the European Planck space probe. The space telescope extended its antennae, cooled with liquid helium, in all directions, in order to determine the temperature of the cosmic background radiation down to a millionth of a degree. From thousands of individual records, the astronomers of many European universities put together a panorama – the celestial map of background radiation.

The map on page 29 shows the beginning of the story. The Earth is still far from having come into existence, but the blotches that look like continents already show where matter has condensed to form clouds of gas, galaxies, stars and, later, planets. Just as ultrasound pictures allow us to marvel at the development of a human being in the womb, read its heartbeat, follow the growth of its organs and eventually make out its features, the development of the universe is revealed in recordings of cosmic background radiation.

Its patterns contain an almost unbelievable wealth of information. When it is analysed, the radiation reveals that ever since it came into being, space has been expanding more and more.

This whole cosmos that we can see today must therefore once have fitted into a tiny volume. The further back we go in time, the smaller it must have been: smaller than the Moon, smaller than a football, smaller than an atom. At some point or other, there must have been a beginning. We call this beginning the Big Bang. In the Big Bang, a tiny universe arose that already contained everything. Since this beginning, nothing has been added to it. The universe has simply expanded and metamorphosed.

That is beyond our imagining. But then, we can't imagine the present universe either. Its dimensions, which can also be worked out from samples of the cosmic background radiation, are beyond anything the human mind is capable of comprehending. The expanses we can see in the night sky may arouse our awe, but beyond the horizon of the most distant galaxies a cosmos that is at least two hundred and fifty times larger is concealed. And that is an extremely cautious estimate. There is good reason to assume that the part of the universe beyond our view is many billion times larger than the segment we can see. There will be more about that in Chapter 9.

What might there be beyond the horizon? The cosmic background radiation that reaches us today marks the edge of the visible world. It took 13.8 billion years, the age of the universe, to reach Earth. The light from areas that are even further away has not had enough time since the beginning of the cosmos to reach us. It will presumably never arrive on Earth: space is expanding too quickly. Today, the cosmos is actually expanding faster than the speed of light. Areas that are far apart are moving away from each other so quickly that not even light can keep up. The celestial map of the Planck space probe therefore reveals that there is a reality beyond the visible universe that is not accessible to us: a world beyond.

But it is precisely to this sphere beyond our vision that we owe our existence. An entirely perceptible universe would be much smaller. But such a universe would have been incapable of creating us. It would have either contained too little material for celestial bodies or would have imploded because of its gravitation long before we humans could have developed.

The celestial chart has even more to tell us about how finely balanced our world is and how little we know about the place where we live. The blotches on the picture of the cosmic background radiation are the seeds of the galaxies. Material has gathered there in order to form heavenly bodies later. What drew the material together was its gravity. But all the mass in the cosmos that is known to us has nowhere near enough gravity to form these patterns!

From the photograph of the Planck space probe, the extent of our ignorance can be worked out right down to the decimal point. 84.5 per cent of matter is 'dark' – a polite way of saying that at the moment, we don't have the least idea what it consists of. The one thing that is certain is that this shadow world exists, just as the hidden spacc does, and also that it brought about our existence, for without it, the gases in the early universe would not have condensed into heavenly bodies. Without dark matter, stars would never have ignited and hatched oxygen and carbon, the basis of life, in their fire.

There seems to be a happy accident hidden in the structures of the celestial chart: the cosmos is huge and therefore stable; it contains dark matter that made the stars light up, and just enough energy to make the cosmos expand without tearing the galaxies and planetary systems apart again as soon as they were formed. Even in its very first light, 13.8 billion years ago, a universe was shining that can bring forth life – us.

'If you're religious, it's like seeing God.' That was what George Smoot said in 1992 about the first pictures of the cosmic background radiation. These photographs, for which the American physicist was later awarded the Nobel Prize, were still extremely blurred; today we can see the first light of the world in all its splendour.

The blue Earth rising over the Moon once showed the people sitting in front of their television sets our planet as an oasis in the uninhabitable vastness of the universe. However abstract it looks, the celestial map of the background radiation, on the other hand, stands for a more benign cosmos that is equipped for life. It's almost as if the cosmos knew we were coming.

3
Riding on a Ray of Light

*A young man wonders what light is,
and his reflections on light explain the world
to him. Time and space are revealed.
But when Albert Einstein dies,
light is still a mystery.*

When I was ten, I was plagued by a recurring dream: the sky darkens and the sun fades. Everything is drained of colour. In the pale light, there is a winding, endless queue of men, women and children. They are all carrying suitcases and waiting at a gate leading into a tunnel. I know that the sun will never rise again once it has gone out, and it's already very dark. My parents take me by the hand, we join the queue and wait, until we too can set off on the path down into the underworld.

Each time I awoke I was distraught. What frightened me about this dream wasn't so much the darkness itself but the idea that it might last for ever, that we would have to live in a world without light. That made me realize how much I was dependent on the Sun. I knew the people I needed, but all I knew about the Sun was that it rises every morning because the Earth rotates.

I didn't find that exactly reassuring. Does the Earth

ever get tired of its movement? And why doesn't the Sun burn out, what makes it shine? And light – what is that really?

My father was a chemist, my mother was a chemist and so were my parents' friends. Sometimes they would take me to the lab and I was allowed to do experiments. They gave me powders with which I could make flames burn blue and green, and thermos flasks full of liquid nitrogen wreathed in swathes of vapour. I felt I was in Wonderland. If I dipped a rubber ball into the freezing liquid and then threw it on the floor, it would shatter. 'Why?' I asked. 'Because the cold turns the ball into ice,' my father told me. He could explain anything.

So I plucked up my courage and asked him what light was. I didn't tell him why I wanted to know, I didn't feel like talking about my dream.

'Light consists of invisible particles,' my father explained, 'the light particles are called photons.' I imagined the photons as raining down on the Earth from the Sun and that with every ray of light a shower of photons was pattering on my skin.

'That's the way it is,' my father said. 'That's why the sun warms us, by sending us energy in the form of particles. Without that energy, we'd just die.'

Many years later I heard the pattering of that shower. I was sitting, somewhat bored, in the main Physics lecture theatre of Munich University, when the professor had a trolley brought in. There was a black tube on the cart, with a dim light and a detector inside it. The photoelectric cell of the detector, the professor explained, was capable of counting photons, one after the other. Every time the detector was hit by a photon, it would make a click. When he switched the machine on all we could hear at first was white noise, but when the professor gradually dimmed the light, there was a brief silence, then it suddenly went 'tock' – as if a drop of water had dripped from a leaky tap. And then again, 'tock, tock, tock…'

So those were the signs of life of the particles that we have to thank for our existence, that give us day and night, the shimmer on water, the sparkle in snow and moonlight, the colours on a soap bubble, rainbows and the flicker of candles in winter.

Light doesn't just transfer energy, however, it also transmits information. It took me another ten years before I really understood that. It was a research project on life in the depths of the ocean that opened my eyes. Some marine biologists had invited me to join them on their research vessel; we were anchored off the coast of California watching on screen the images that a robotic submarine was sending up from thousands of metres below. We spent most of our time staring at black monitors, but occasionally a few flashes would skip across the screen and then we would cheer – an inhabitant of the deep sea had announced its presence. I suddenly realized that this world is not just cold, but also without knowledge, that in the eternal darkness no intelligence can develop. The creatures down there that feed on the crumbs that come down from the layers of the ocean illuminated by the sun do not know what it is that surrounds them. Angler fish and krill, vampire squid and lantern fish have all developed luminescent organs so that they can communicate with each other, but the flashes of light, produced by luminescent bacteria living in their bodies, are too weak to light up even their immediate surroundings. An awareness that their habitat is only a tiny part of the world remains inaccessible to the inhabitants of the ocean depths. How could the angler fish ever find out about the star-studded sky that arches over the sea?

We humans also only perceive a tiny snippet of reality. For most of our history, our ancestors thought their planet was the centre of the universe. It took thousands of generations before they began to see Earth as one planet in the cosmos. However, unlike deep-sea fish, we are capable of asking questions. Even as children we want to find out what is around us, where we belong. We suspect that we are part of something bigger, that we belong to a reality that, at best, we only know in part. These questions catapult us out beyond the narrow confines of what we can perceive around us.

More than a hundred years ago, a schoolboy asked whether we had to understand light in order to understand the world. Perhaps the only person who is able to think in this way is someone who has not yet had time to form strong opinions and prejudices.

There was certainly reason enough for that boy to think about light, for that was how his father earned a living. In 1886, the Munich electrical engineering firm of Einstein & Co. was awarded the contract to illuminate the Oktoberfest for the first time. A report in the Oktoberfest newspaper went into raptures about how 'the soft and yet so intense glow of the electric arc lamps cast a magical light over the thousands of people at the festival'. Three years later, the firm was making plans to light the streets of the Schwabing district. Albert Einstein was helping out in his father's firm, explaining to his classmates how a telephone worked and reading popular-science books. At 15, he wrote his first scientific article. It was about the spread of light in empty space. He wondered what it would be like to run after light. If an incredibly fast sprinter managed to reach the speed of a sunbeam, would he be able to capture the light? And what would it be like to ride on a ray of light?

This question occupied him for 11 years. In the summer of 1905, he finally found the solution. He sent two articles to the most prestigious scientific journal, the *Annalen der Physik*. In the first article Albert Einstein showed what light is; in the second he explained how we can understand the universe if we can understand light.

These two articles changed the world. The first examined light as energy. It was from this that the quantum theory, the physics of the tiniest units, was developed. It is about a remarkable world that is dealt with in Chapter 5 of this book. In this world, nature can make leaps and bounds, particles go through walls and things happen for no apparent reason. In the second article, Einstein looked at light as information. That formed the basis of his theory of relativity, which provided a new explanation of space and time.

So it was that in that summer of 1905, mankind's horizon was expanded. It opened up in two directions: the quantum theory was intended to show how the world works on the smallest scale, what all things consist of, the theory of relativity the kind of cosmos we live in. It was as if the fish of the ocean depths had suddenly realized they were swimming in water, that the ocean had shores and that there was land beyond the coast. (And if that wasn't enough, Einstein published a further article in the *Annalen der Physik* in that same summer. In it, he proved what many scientists were still denying in 1905: the existence of atoms.)

How could one man of 26, working as a third-class technical expert in the Bern patent office because no university would hire him after he had completed his physics degree, get so far?

Einstein's starting point was the question that had already been bothering him when he was a schoolboy: how does light move? At that time, it was already known that the speed of light, an incredible 300,000km (186,000 miles) per second, does not depend on whether the source of light is moving. There are puzzles that can be solved step by step when you think about them, and others that seem ever more unfathomable the more you consider them. For Einstein's contemporaries one such mystery was the propagation of light.

Isaac Newton, the father of modern physics, had explained that light consists of tiny corpuscles of material. But things come closer more quickly if you're running towards them, and more slowly if you're running away from them. All matter obeys this simple law of motion. It corresponds to the experience of the world on which Newtonian mechanics is based. Light, on the other hand, always moves at the same speed, meaning that it cannot consist of corpuscles. Light is not matter.

Once Newton's theory of light corpuscles had failed, people adopted another explanation: light was like a wave. Waves are oscillations in matter. The swell spreads across the ocean, sound in the air. But what does light oscillate in? Clearly it spreads in a vacuum, otherwise the rays of the sun could never reach us. However much time the physicists of the 19th century spent looking, they could find no medium in which light could oscillate. Light, therefore, is not at all like a wave in the ocean. But it doesn't behave like matter either. What then? That was the situation when Einstein started thinking.

Einstein was undeterred. He relied on his imagination. 'Imagination is more important than knowledge. For knowledge is limited to all we now know and understand, while imagination embraces the entire world,' he told an interviewer who had asked him how he had arrived at his discoveries. Could light, for example, consist of weightless articles? Particles without mass – it sounded absolutely crazy. But the hypothesis solved all the problems. Because particles without mass have no inertia, they can be neither slowed down nor speeded up. Therefore light always moves at the same speed. Moreover, the idea of massless particles explains what happens when light hits an object. In such cases sensitive measuring devices, as Einstein knew, report impacts, as if the surface were being bombarded by tiny projectiles – the impacts I had heard in the Physics lecture: tock, tock, tock…

But however attractive his ideas sounded, Einstein was still unsure. He preferred to express himself very cautiously. Radiation, he wrote in that summer of 1905, behaved 'as if it consisted of independent quanta of energy'. But that said everything: light is pure energy. This energy is bundled up in tiny packets – photons. They are the particles without mass. With this theory Einstein could explain all the measurements.

So had he explained the mystery with that? The words 'as if' are there as a warning. No one should think they have now understood the true nature of light. The reader has just come a good deal closer to the secret. But everything we can imagine is always permeated with our experience of everyday life. Even in our most improbable dreams, we see ourselves surrounded by things we can look at and touch. However, the world into which Einstein takes us has a quite different, unfamiliar form. In it there are energy quanta that have no mass and not even spatial extent, but still behave like particles.

Einstein was not prepared to simply accept comfortable ideas. He was therefore constantly testing his theories out in new 'thought experiments', which were confirmed. Doubts that came close to self-torture led him to his discoveries.

In 1922 he received the Nobel Prize, but it was awarded for only a small part of his work. Einstein was awarded the honour for the formula establishing the relationship between the energy and the frequency of oscillation of light. However, it was specifically stated that this highest scientific award was not given for the insight that allowed him to realize the existence of photons. Einstein's imagination was too much for the Royal Swedish Academy. And the committee were at pains to make it clear that the prize was not being awarded for the theory of relativity, with which Einstein had provided a new explanation for time and space. That a person travelling quickly is ageing more slowly, that light can bend space, that past and future lie in the eye of the beholder, seemed too fantastic for the majority of the members of the Stockholm committee to be worthy of the prize. Perhaps they also found the theory of relativity too disturbing. Einstein was even instructed not to mention it in his speech of thanks. In fact by the time he heard about the impending honour, he was already on board a ship, bound for a lecture tour of Japan, and thus never accepted his Nobel Prize in person.

Yet all Einstein had done was to take his thoughts about light to their logical conclusion. The search for an explanation for the energy of light had led him to photons. The theory of relativity, by contrast, describes light as a carrier of information. The principle is very simple: since light moves at a finite speed, we only see part of the world. We only learn about events whose light reaches us. Therefore light determines our knowledge of the world.

This doesn't bother us much in our everyday life, because light spreads very quickly and here on Earth we don't look particularly far anyway. But this vision is similar to the perspective of the deep-sea fish that can hardly perceive its immediate surroundings. If, however, we look up at the sky, then the speed of light certainly does play a part. Light from the Moon takes one second to reach us, from the Sun eight minutes, from the most distant planet, Neptune, four hours. When you're looking at far-away objects, you're looking at the past.

And there's more: light determines what we see as past, present and future. That is contrary to our intuition. Our view sees time as the basis of everything, therefore we imagine that time cannot be dependent on anything else, that it's an absolute. And we think that the present, past and future mean the same everywhere. But our vision also tells us that the Earth is flat. Einstein, however, subordinated intuition to a law of nature. This law of nature is the constancy of the speed of light. Einstein realized that we are dependent on light for everything we can say about space and time. Consequently, in our understanding of past and future, we have to take account of the fact that light happens to be moving at a finite speed. Why should laws of nature do us the favour of fulfilling the expectations of our intuition?

Imagine two people, Alice and Bob. If Alice were to observe two flashes of lightning from the same distance at the same moment, she'd say that the two flashes appeared at the same time. But Bob, who is moving quickly past Alice, perceives the event quite differently. He's flying towards the one and away from the other, therefore the light from the flash he's approaching will reach him sooner than the light from the flash he's moving away from. So he sees one flash sooner, the other later. What Alice experiences 'simultaneously' is in Bob's eyes a 'before' and an 'after'. At the point when Alice is perceiving both flashes, he has already seen one and is still to see the other. What is the present for her, he calls past and future. Einstein realized that there was no reason why just one of them should be right. All that is left for us is to say farewell to one time for all. Every one of us lives according to their own time.

The fact that people age more slowly when travelling became famous. This affront to our intuition can also be explained by the properties of light. We don't perceive this effect at the speeds we normally move at, of course. But it can be observed in certain small elementary particles, called muons. Normally they decay within a millionth of a second, but when they plunge down onto the Earth at great speed, they survive longer. There can therefore be no doubt the lives of humans would become longer if they were to board a very fast spaceship.

The slowing down of time during fast motion finally provided Einstein with the solution to the puzzle that bothered him as a schoolboy: what would it be like to take a ride on a ray of light? The greater the speed, the slower the clocks go; when the speed of light is reached they would stop. For light, time is standing still. The rider on the ray of light would live in an eternal present.

But the theory of relativity rules out that possibility. Einstein only realized why after he had completed his three great publications of the summer of 1905. Again, he didn't feel sure of his conclusion. 'It's an amusing and seductive idea, but I just don't know whether the Lord God is pulling my leg and having a good laugh about it,' he wrote to a friend. Nevertheless, he sent an addendum to the *Annalen der Physik*. On just three pages, he demonstrated that from the invariance of the speed of light it follows that a body will increase in mass when it is subjected to acceleration. It is this extra mass that contains the energy of the acceleration. If, however, the speed should approach that of light, the mass of the body will increase disproportionately. It will just become heavier, not faster. That is why the object will never reach the speed of light: the energy that is used for its acceleration turns into mass. At the end of his short calculation, Einstein explained the conclusion. One single sentence, itself like a beacon: 'The mass of a body is a measure of its energy content.'

Mass is energy. $E = mc^2$. That is the formula as we know it today; Einstein wrote it down as $E = mV^2$, because 'c' was not yet in use as a symbol for the speed of light at the time. This relationship made the atom bomb possible. But the whole of modern cosmology also rests on it. Einstein was to spend a further ten years thinking about the consequences of the equivalence of mass and energy, finally publishing his masterpiece, the general theory of relativity, in the autumn of 1915. In it he explained how energy, alias mass, determines time and space.

A mere four years later this construct of ideas was confirmed. And once more it was light that played the decisive role: astronomers used the total solar eclipse of 29 May 1919 to prove that visible stars close to the Sun appeared displaced compared to their normal position in the sky. This observation corresponded precisely to Einstein's predictions: the mass of the Sun distorts space and thus diverts light from the stars. Normally this effect cannot be seen because the Sun outshines the weaker starlight. However, when the Moon darkens the sun, the distortion is noticeable.

Einstein managed to bring together his whole theory of relativity in a single formula. On the one side of the equation are energy and matter, on the other space, time and their distortion. These quantities are linked together by light. It was with this equation that Einstein and his colleagues started their calculations. The result threw all the ideas people had ever had about their cosmos into disarray. Einstein's theory led to what appeared to be the most abstruse ideas: the universe is expanding; clouds of gas contract, igniting stars in the heat of which elements arise; dead stars implode and, as black holes, swallow up light and matter; gravitational waves make space pulsate.

Einstein himself scarcely believed that one day every single prediction of his theory would turn out to be true. A 1916 report from the Prussian Academy of Science, for example, reveals how cautious and at the same time prescient he was: Einstein proved to his gathered colleagues that gravitational waves had to exist – but with the reservation that these rhythmic vibrations of space were too weak ever to be measured. And, indeed, it was to be a hundred years before American physicists published the first recording of gravitational waves. The signal that was discovered in 2016 recorded the collision of two black holes a billion light years away from the Earth.

His reflections on light enabled Einstein to see a world in which space and time, mass and energy were no longer independent factors. What one observer sees as simultaneous, appears to another as earlier or later. And the present for the one is already the past or the future for the other. Masses bend space, matter turns into energy.

In such a world, it's not things that count, but events. Energy and information are what matter, not tangible objects. In such a world, computers, solar cells, satellite navigation and the Internet can exist, but also nuclear reactors and hydrogen bombs. Each one of these inventions derives directly from Einstein's discoveries about light.

The questions he asked himself as a schoolboy never loosened their grip on Einstein. Right up to the end of his life he couldn't accept that massless light particles had properties that went beyond even his imagination. In 1951, a few years before his death, he wrote to Michele Besso, a childhood friend: 'All these years of conscious pondering have not brought me any closer to answering the question, "What are light quanta?" Nowadays any Tom, Dick or Harry thinks he knows, but he's wrong.'

The way to understanding is not a straight, flat

road but a spiral staircase. Anyone who turns around a number of times while covering the distance ends up looking in the same direction as when they started. But they will have reached a higher level.

4

The World Spirit Fails

*A hurricane sweeps across Germany,
a storm no one saw coming.
Reasons why the world is unpredictable,
and praise of the creative universe.*

We have telephone conversations with machines that talk, let navigation devices guide us through strange neighbourhoods and will soon entrust our lives to self-driving cars. I am amazed at how friends whose first marriages to partners they chose for themselves failed now find happiness with men and women a computer selected for them. Not that I have any objection in principle to machines influencing my decisions. Apps that promise to accurately predict the weather to the hour, fourteen days in advance, have taken up residence on my mobile phone. At the start of the week, a sun icon reassures me that the garden party planned for 4 o'clock on Saturday afternoon can go ahead, while a rain cloud symbol makes me doubtful about the canoe outing we've already arranged for Sunday morning.

Meteorologists can be proud of how much they've achieved. In 17th-century France, people who foretold the weather were bound to the wheel as fraudsters. Bismarck banned the introduction of a state weather service on the grounds that a Prussian official could never be wrong. Today, meteorologists base a multi-million-pound business on being able to predict the future ever more precisely.

The weather predicted today for tomorrow occurs with a probability of over 70 per cent; nowadays the forecast for the next three days is more reliable than that for the next day was when I was a student. This is, of course, all thanks to computers. The supercomputer Cray-1, cooled with liquid freon, which in 1979 was the herald of a new era at the European Centre for Medium-Range Weather Forecasts, could perform 100 million operations per second. Today the Apple Watch can calculate thirty times faster. The current machines of the weather service are so powerful that they can resolve the sky over Germany down to just 3km (a little under 2 miles) – a forecast for every village. To achieve this, the computers devour a stream of data from more than ten thousand weather stations, thousands of aeroplanes and ships, and dozens of satellites.

The world has been scanned, captured and translated into machine-readable information: we have come remarkably close to the dream of Pierre-Simon Marquis de Laplace. In 1814 this Parisian astronomer dreamt of a perfect intellect that was 'vast enough to submit all data to analysis'. Nothing would remain hidden from such a 'world spirit', known as Laplace's Demon. However, what Laplace had in mind was not simply global surveillance; rather, he wrote, his perfect intellect would be able to foresee the future. For everything that happened, he went on, followed the laws of nature, and anyone who knew the present state of the world precisely enough would be able to calculate the situation for the next day. For Laplace these reflections were just an intellectual game. No one could imagine that the mass analysis of data would ever become a reality. The Demon was a metaphysical being, thought up in order to demonstrate the power of the laws of nature and to overthrow a God exercising His own free will. Laplace's contemporaries could either follow him in believing that the world is predictable, or not.

To us, however, this vision doesn't seem at all fantastic any more. Less than eighty years after the inventor Konrad Zuse put the first programmable computer into operation in Berlin-Kreuzberg, we can analyse as much data as we like. You just have to set up enough machines. The Demon has found its home in the processors and since their capacity doubles every eighteen months, there seems to be nothing to stop us from seeing the future with ever greater accuracy.

Internet companies and secret services, which already have huge computing capacity at their disposal, want to be able to predict human behaviour down to the very last detail. They would like to know what kind of advertisements will ensnare us and where the next terrorist attack might take place. Doctors hope that genetic information will make illnesses predictable, while neuroscientists and some philosophers even consider the human brain a suitable case for Laplace's Demon. They dream of a mathematical model for everything that goes on inside us. Is it surprising, then, that so many people are frightened of research? Behind this anxiety lies the worry that the world might lose its magic – and people their freedom. We want to be unpredictable.

The idea that everything that happens in the world can be attributed to just a few laws and can therefore be calculated, in principle at least, is known as reductionism. Reductionism is something like the creed of modern science. Research subsists on finding explanations that are as simple and at the same time all-encompassing as possible. We can fly to the Moon because we can deduce the movements of all bodies in space from one single law of gravity. We have explained the development of all living beings from Darwin's laws of mutation and selection. As a result of understanding the dynamics of the atom through the equations of quantum physics, we have been able to create a world full of computers and lasers. Successes like these have allowed our faith in reductionism to grow to immense proportions.

Given all that, who cares that we can't even predict the behaviour of an ant on the basis of the laws of nature? Anyone who believes in reductionism simply takes refuge in their lack of detailed knowledge. We are a long way away from understanding the brain of an ant, we don't even know how many types of neuron are firing inside it. But the more we learn, the greater the progress we will make in predicting ant life – at least that is what we hope.

Things are much easier for meteorologists. There are no mysterious events in the atmosphere. It has become impossible for us to believe in rain spirits or thunder gods; we know that ice crystals and tiny drops of water gather in the clouds. All the different types of weather conditions can be explained by the fact that the Sun warms the Earth. That is why water vapour and dry air are set in motion in the atmosphere, gases expand and vapour condenses into little drops and freezes to ice. All that is elementary physics. When the air warms up, it has to become thinner. When air becomes thinner, it has to rise. When air rises, new air from the ground has to flow after it – a wind blows. Air and vapour consist of molecules and the dynamics of the molecules is determined by an equation that can be written out in a single line. This formula, the Schrödinger equation, is the formula for everyday events throughout the world. All matter obeys it, at least under the conditions normally found on Earth. It determines the way crystals are formed, the way flowers and trees grow, and the way messenger substances in the brain operate – and the weather, of course. The Schrödinger equation contains nothing but the masses and electric charges of atomic nuclei and electrons, the intervals between them, and also a natural constant. Despite that, its accuracy is beyond all doubt: measurements have confirmed it down to at least 18 decimal places. So what is stopping us from getting supercomputers to work out what the weather will be like in three weeks' time or in the summer after next?

Laplace's Demon gets its chance every evening after the news on television. But instead of the expected sunshine, a summer storm with heavy rain arrives on the afternoon of the garden party. Or conversely, more than 65 per cent of storms forecast by the German Weather Service never happen. And what private meteorological companies offer in terms of forecasts for more than a week ahead is rarely much better than a roll of the dice. Despite the fact that everything happens according to simple and well-known rules, the forecast fails.

Is it all a bit too much for the Demon? This problem reminds me of playing chess with my son.

When he was five, he wanted to learn how to play and within half an hour, he had a grasp of the rules. Now he knew how castles, knights and bishops move and take pieces, that he could exchange a pawn for a queen if it reached the opposing back row and that he had to checkmate me. Elias had even understood castling. But to his great disappointment that wasn't enough to beat his father straight away. His moves led nowhere. He knew the rules that determine everything, but he hadn't understood chess yet. He lacked the vision and the experience a chess player develops over many years. Knowing the rules alone didn't mean Elias could make a meaningful move. No one can do that. You have to play chess and explore its possibilities in thousands of games.

In the early days of computers, Claude Shannon, the father of information theory, had the idea of teaching machines to play chess. Laplace's Demon would be unbeatable at chess; with its perfect intelligence, it would be able to work out every possible course the game could take, and would therefore always make the best possible move. However, in every position chess players have the choice of around 30 legal moves and in a typical game, black and white move about 40 times each. Therefore we get the huge number of 10^{120} possible courses the game might take – that is a one with 120 zeros. In order to go through all possible options, even one of today's supercomputers would have to spend 10^{90} years calculating – that is many billions upon billions times longer than the world has existed since the Big Bang. That is how much time would pass before an unbeatable chess computer made its first move.

These are numbers that are beyond our comprehension. It's best to call them by a name made up by a child. This word was invented in 1938 by the nine-year-old nephew of the American mathematician Edward Kasner. He is said to have asked the boy, while they were out on a walk together, to come up with a name for the number 10^{100} – a one with 100 zeros. It's the number you get when you multiply ten billion by itself ten times. 'Googol,' the boy replied without hesitation.

That 'Googol' sounds like the name of a search engine is of course not mere chance. In 1997 some PhD students at Stanford University remembered the word the boy had coined when they were looking for a catchy name for a new website that would make immense amounts of information accessible. As the story goes in Silicon Valley, spelling was not exactly the strong suit of the young computer scientist who typed the word into his terminal.

The number of possible variations in a game of chess has a few zeros more than a googol, the time in years a perfect chess computer would need to think them through a few less. But with numbers of such magnitude a few zeros make no difference. At any rate, these are numbers we can't handle. The googol denotes a transition to infinity. It is still a normal number, but one so gigantic that any calculation with that many possibilities is out of the question. We will be encountering the googol again.

We hope in vain for better technology. For example, that chess computer would have to store all the data it creates somewhere. But a memory large enough for an unbeatable chess programme cannot be built. To store information, you need material – a piece of paper, a brain, a computer chip. And because every material object consists of particles and therefore cannot be written on infinitely closely, the capacity for storing data is always limited. The most extensive memory imaginable is the visible universe – the part of the cosmos whose light reaches us. It so happens that the amount of data that all the particles in the universe can accommodate is roughly the same as the number of possible variations in playing chess.

That is why Laplace's Demon cannot play chess, that is why there will never be a perfect chess computer. The existing programmes are more modest in their aims. Instead of calculating all possible permutations, they just examine the most promising moves. That is much quicker. However, they do need prior knowledge about what usually works in chess. So the machine is programmed to deal with known positions, with openings and endgames, or to take into account the fact that exchanging your queen for a knight rarely pays off, which is something I also explained to my son. The computer and the child are no longer selecting moves solely according to the rules of the game but also on the basis of things learnt from experience that they have been taught. So the programme is foregoing a secure prognosis and making do with probabilities. That makes it vulnerable, because in some situations experience can lead us astray. Chess computers that defeat world champions are the one-eyed men among the blind. These computers cannot provide perfect prognoses, it's just that the human mind is even less able to grasp the complexity of the game than they are.

Compared with the weather, chess is straightforward: the game consists of only 32 pieces and 64 squares, and those already allow for billions of permutations. But there is immeasurably more going on in the atmosphere. One single breath, a quarter litre of air, contains roughly 10^{22} oxygen and nitrogen molecules. And the whole atmosphere has a volume of about 10^{22} litres. (That is a one with 22 zeros. Enrico Fermi, who in 1942 achieved the first controlled nuclear chain reaction, is said to have conveyed these figures in an amusing way: more or less every time we inhale, we take in oxygen molecules from Julius Caesar's last breath. Over two millennia, the air the dictator exhaled at the moment of his death has spread evenly over the whole Earth. That is to say that on average, every litre of air contains one molecule from Caesar's last breath.)

No brain, no computer in the world will ever be capable of reproducing these particle dynamics. And if chess programmes are unable to try out all the positions on that little board, then it will certainly be impossible to work out all potential atmospheric conditions. So we're stuck in a remarkable dilemma: on the one hand the weather is nothing more than the movement of air and water vapour according to elementary laws. It is to these dynamics alone that we owe snow and gales, bright blue skies and rain. On the other hand, these simple rules that everything obeys don't show us the storm front. If we wanted to get the fastest digital computers to derive the shape of a nice little cotton-wool cloud from these rules, the machines would be occupied for several times the age of the universe and would produce more data than the cosmos can take. If you study nothing but the elementary laws of nature, you will never find out anything about hailstorms, monsoons and tornadoes, sleet and spells of fine weather. The diversity of nature is only revealed to you when you see it.

If the whole universe as a computer is not big enough to draw useful conclusions from the basic facts, then reductionism is a paper tiger. The only way out is to change our perspective and speculate about clouds, winds and areas of low pressure.

Thus the world is beyond our knowing. But if we abandon the pedantic approach of the Demon, we can see patterns in our immediate surroundings. After all, matter loves to get organized: molecules come together, clouds form in the sky. Therefore wanting to predict the future is not a hopeless venture. We should model ourselves on chess computers or rules of thumb, both of which ignore details and just follow the main lines. Often the details aren't crucial, in which case an out-of-focus view is better. Of course, the most we can say in that case is what will *probably* happen, but with simple processes this probability is very high. In order to venture a prediction about the temperature at which water will boil, you don't need to know how every molecule in the kettle moves.

But which patterns influence the weather? Structures of all conceivable sizes can be identified in the atmosphere. Microscopic swirls fill every thimbleful of air. Floating in the cumulus clouds that promise fine weather are tiny drops of water less than a thousandth of a millimetre in circumference. The cloud itself, however, trails on for 100m (325 feet).

A thundercloud in the sky spreads out over 10km (6¼ miles), a low-pressure vortex over 1,000km (620 miles). The largest formations, for example the vast convolutions of the jet-stream, encircle the Earth and can lock in air masses over a whole continent.

And every structure, from the smallest to the largest, has its own role to play. In the turbulent flow of a cloud tiny drops of water collide and combine into visible drops until it starts to rain. And then the cooling of the ground after the rain can bring about a change in the weather because the atmosphere is unstable. A mere breath of wind can disturb it. One disturbance sets off another, like a single toppling tile knocking over an entire line of dominoes. And because the Earth is rotating, such an event makes an impact over huge distances. The weather over Germany is often as good as unpredictable if there have been violent storms over North America a few days previously.

That even the flapping of a butterfly's wings on a distant continent can bring about a change in our weather has become a modern adage. The American meteorologist Edward Lorenz coined it in a 1972 lecture about the difficulties of weather forecasting. He wasn't saying that the world was not subject to any rules, but that it was impossible to know all the decisive influences on a forecast. We can't keep our eye on all the butterflies in the world.

A friend of mine experienced the butterfly effect on Boxing Day in 1999, when she tried to cross the Swabian Alps in her convertible. That morning's weather forecast had confirmed there was no chance of snow or other storms, but on the Stuttgart-Munich Autobahn Ulrike witnessed apocalyptic scenes: trees falling on moving cars, HGV trailers flying through the air. She had unexpectedly ended up in one of the worst hurricanes ever recorded in Europe.

The drama had started on Christmas Eve on Sable Island, a sandbank off the coast of Canada also known as the 'graveyard of the Atlantic' because of its many shipwrecks. The sandbank is uninhabited; the only people out there are the five employees of a weather station. They sent up a sounding balloon on 24 December 1999 to measure the temperature, pressure and wind speed in the upper reaches of the atmosphere. But the probe failed and the flight was curtailed. A little later, the machines in the European weather centres started their computations; all they had from Sable Island were the few data the balloon had managed to send back down to Earth. This forecast predicted severe storms over south Germany for Boxing Day, but was never released.

The reason was that a second probe was sent up from Sable Island exactly 114 minutes after the failed first balloon. This time the measuring instruments worked, however the technicians forgot to reset the probe's clock to the later start. That meant that the time signal transmitted back to Earth was almost two hours slow, but the meteorologists at the German Weather Service didn't notice. They therefore assumed that the wind high above Sable Island was 20km (12 miles) an hour slower than it had been in the previous measurement. This discrepancy seems unimportant, given that the upper winds can reach up to 500km/h (300mph), and anyway, the figures from Sable Island are only one tiny detail in the greater overall picture of weather activity worldwide that is compiled from thousands of data sources. However, the computers did the calculations again using the minimally changed base data, but this time the outcome was quite different: there was no longer any mention of a storm. And this result seemed more plausible to the meteorologists on duty. On Christmas Eve the television news promised calm days with a 'typical Christmas thaw'.

In the meantime, south of the Azores a previously unremarkable shallow area of low pressure was turning into a little hurricane vortex. This development also went unnoticed. It was the turn of the meteorologists of the Free University of Berlin to name the depression, and they chose to call it 'Lothar'. It was not until the evening news on Christmas Day that isolated, though harmless squalls were announced for North Germany. At that point the hurricane was already hurtling towards Brittany. At four in the morning on the next day it hit the French coast close to Brest. In south-west Germany the air pressure dropped at a speed never before recorded in Central Europe. Thus Lothar turned out to be a 'meteorological bomb' – that is the technical term for a low-pressure vortex that develops with great intensity. At around eight o'clock the hurricane tore a clock tower off the cathedral in Rouen. Gusts of over 170km/h (105mph) were tearing through the streets of Paris, lifting the 1500 tons of the Pantheon's lead roof. Then the front crossed the Belgian border, reaching North Baden and Hesse around midday. The wind speed continued to rise. In the Black Forest a diesel locomotive was whirled off the rails. In Baden-Württemberg alone, where my friend was sitting out the storm in her red convertible, more than 50 million trees fell down. Sixty people were killed. In the late afternoon the storm gradually subsided over the Czech Republic.

Could this disaster have been foreseen? The sloppy work of the Canadian meteorologists on Sable Island is only a superficial explanation for the catastrophe. If the second probe had transmitted the correct time, the weather services would perhaps have been in a position to issue a warning about the storm one day earlier. Then again, this single inaccurate piece of information among tens of thousands of correct figures was enough to make the whole forecast useless. Thus the Lothar disaster shows why all hope of a reliable weather forecast is in vain. What happened over Christmas 1999 is bound to happen again: even if all the systems function perfectly, that kind of inaccuracy cannot be entirely avoided. Every measurement contains mistakes. It is not even theoretically possible to determine events in the atmosphere precisely because the whole universe does not possess sufficient computing capacity. Even if it was just a shower that Laplace's Demon wanted to predict, the cosmos would be too small for him.

Thus the dilemma is unavoidable: we cannot process all the data, but incomplete data means a perfect forecast is not possible. All we can do is to live with the imprecision and reduce it. Faster computers and better measuring instruments give us an increasingly precise view and can at least improve the forecast for the first few days. But then the butterfly effect comes in again. The mistakes increase with every step in the calculation and soon run rife again, making any further forecast impossible. The bank of cloud blocking our view of the future has only shifted a little further back.

Storms will continue to take our descendants by surprise even after part of humanity has already left for other planets. In the end our understanding, no matter how many computers it has backing it up, has to capitulate when faced with reality. It is not complex correlations but the immense numbers of the simplest particles that defeat every intelligence.

And the twists and turns of love, financial crises and even the seething activity of ants in the anthill really will remain forever incalculable. Just a brief look inside our heads would make Laplace's Demon dizzy. The human brain contains 10^{14} synapses, that is a thousand times more connections between brain cells than there are stars shining in the entire Milky Way. How could the World Spirit ever predict the prospect of a romantic evening or foretell a marital crisis from the state of our grey matter?

Nature allows us deep insights into its regulatory system while at the same time preventing us from seeing what it's really up to. Some scientists might regret that, others among us will be relieved to note that our emotional life remains incalculable.

There is another reason the failure of the demon makes me happy: it reminds me how wonderfully nature is constructed. Everything around us is made up of atoms. The atoms are tiny and extremely numerous. Incredibly complex structures can arise from a huge number of simple building blocks. In clouds, whirlwinds and brains we find nothing but atoms. However, by combining according to simple rules, atoms can produce new phenomena: the weather, our thoughts, love. Our forecasts fail because the universe is creative.

The English physicist Paul Davies once pointed out a remarkable characteristic of the molecules on which all life is based. The proteins and DNA that carry genetic information are chains formed out of a few hundred to a few million building blocks. With proteins, composition determines their form and therefore their function, with the DNA the code of the genetic information. But it is impossible to predict all potential protein forms, all genetic combinations. The chains are so long that once again the whole universe as a computer would not be enough. It is as if nature had ensured that it would keep on surprising itself. To me it seems that it is precisely this unpredictability that marks the boundary between life and death.

5
A Crime Story

A villainous gang is raiding flats in
London and New York.
Although the burglars were not able to arrange things
with each other, their raids are perfectly coordinated.
Investigator Glock is looking for a secret plan,
but cannot find one. His conclusion:
all the places in the world are in reality one place.

*Naturally, we were all there — old Qfwfq said —
where else could we have been? Nobody knew then that there
could be space. Or time either: what use did we have
for time, packed in there like sardines?*
ITALO CALVINO, COSMICOMICS

*I think I can safely say
that nobody understands quantum mechanics.*
RICHARD FEYNMAN

Glock had long suspected that all the places in the world were just one single place. He found the proof during one of those November weeks when the fog hung so heavily over London that you could hardly see the houses on the other side of the street. On days like those, Glock normally didn't go out at all. As soon as he'd had his morning cup of tea, he would turn to his manuscript, which he mockingly called his shackle. After a short time, Glock got so caught up in his work that he sat there almost motionless; he didn't even look up and only left his desk when he felt the pangs of hunger or thirst. He never volunteered any information about what he was working on, but if someone asked him, he would say that it was about a world beyond space and time, which we lived in without noticing it. There was, he said, no difference between 'here' and 'there', but it was still too early for him to explain what lay behind his thoughts. Glock had been polishing his manuscript for years, without getting any closer to publication, but that hardly seemed to bother him at all. His fees and his reputation as an investigator of economic crimes allowed him to take on only the most lucrative and interesting assignments. The remaining time he devoted to his physico-philosophical studies.

Without Glock noticing, it was now late afternoon and the cars already had their headlights on. He had just got up to pour himself a whisky when the doorbell rang. He went to open it. Standing there was a slim woman, perhaps thirty-five, wearing a beige cashmere coat. Dangling from her shoulder was a handbag of brown and doubtless very expensive leather that was folded like a piece of Japanese origami. Her eyes darted around, as if in search of something to hold on to.

'You are Alice Aspect,' Glock said. 'I've been expecting you. Please do come in. I'm John Glock.'

'Thank you for making time for me at such short notice. I'm afraid you're the only person who can help me.' Aspect spoke fast, with a barely noticeable French accent; she had clearly been living in London for a long time.

Without a word Glock showed her into his study. The room was furnished with Chinese antiques, sparsely but to great effect; on a dark burr-wood dining table were stacks of books and magazine clippings. Glock pulled out a chair for his visitor.

'What can I do for you?'

'They arrested my husband the day before yesterday. Bob's now being held on remand. I consulted a defence lawyer immediately, of course, but he told me he thought the case was as good as hopeless – if my husband is lucky, he'll get away with five years in prison.'

'What is he accused of?'

'Gang robbery and forming a criminal organization. You may know that my husband is a partner in one of the world's largest auction houses, Aspect and Associates. They have salesrooms in New York, Paris, Geneva and Vienna. The head office is here in London. Allegedly, rugs and jewels from burglaries have been coming under the hammer in our London and New York branches for years.'

'If your husband knew about it, he is guilty of receiving stolen goods, at most.'

'I promise you, he didn't know. But the amazing thing is how the break-ins are supposed to have taken place. All the jewels and rugs that the investigators seized in our New York branch were stolen in the SoHo district of Manhattan. And without exception, the stolen goods that my husband had unsuspectingly put up for auction here came from Soho in London. That gave Scotland Yard and the New York Police Department the idea of comparing their records. And here it comes: if there was a break-in one night here, one was carried out over there as well. And every time the thieves were after the same things. If jewels disappeared in SoHo in New York, then someone in our Soho lost their jewellery as well. But if the crooks over there took rugs, then rugs were stolen here the same night too. As if our Soho were a mirror image of SoHo in New York. Or vice versa.'

'Or as if the burglars were working together.'

'Quite. But Scotland Yard say that they can rule out exactly that. They refused to tell our lawyer why. Scotland Yard think that someone must have been planning and coordinating the raids well in advance. And according to them, my husband is that someone.' Glock thought he could hear desperation in her voice.

'Truly an astonishing case,' he replied. 'You'll be hearing from me.'

The next morning Glock strode through the mist to Victoria Street. He passed the revolving metal prism bearing the words 'New Scotland Yard' and asked for Inspector Stone at reception. Bert Stone sat in a tiny office on the 18th floor, where Glock had also spent many years of his life. The desk along one wall of the room left just enough space for a chair for visitors. On the wall behind Stone was a street map of central London and another of Manhattan.

'You know why I'm here?'

'I can well imagine,' Stone said. 'We've been keeping track of these break-ins for years. This whole thing is completely crazy. Several times a week a police station in Soho reports that jewels or rugs have gone missing somewhere. But always just one or the other. Apart from that, we could find no pattern, no clues, nothing. All we know is that jewels and rugs disappear with exactly the same frequency.'

'As if the burglars were tossing a coin to decide what to take. Heads it's a rug, tails a few jewels.'

'So you might think. We've taken a good look at the burglaries in Soho. Our database records whether the burglar got in through a door or a window, for example. Or whether the break-in was on the ground floor or one of the higher floors. And whether they took a rug or jewellery. It turned out that the burglars went in through the door just as often as through a window, targeted ground floors just as often as the higher floors.'

'Do you know anything about when they do what?'

'That's just it. We thought, for example, that in some streets they might prefer the ground floor, in others the flats higher up. Or one on weekdays, the other at weekends. But nothing of the sort. We searched the database for all possible patterns but we found no regularities, no system, absolutely nothing to explain the break-ins.'

'You're dealing with a very smart gang. They toss a

coin – heads it's rugs, tails it's jewels. Then they throw dice to see whether they're going to break in through the door or the window and on which floor. And they only make this random decision right before they're going to commit the crime, so the police will never be able to predict where they're going to strike next. It's the perfect strategy to drive you crazy.'

'You can't be serious,' Stone said impatiently. 'These people are professionals. They don't leave anything to chance.'

'I'm completely serious,' Glock replied calmly. 'Your Soho crooks are simply refusing to stick to what you think are good reasons. Because they're acting purely according to chance, they're way ahead of you. Am I right in thinking that you've been trying to catch these guys for years? And you haven't caught a single one so far.'

'Fine. So you believe in crooks who throw dice and I believe in crimes that follow a regular pattern,' Stone said. 'But why are you deigning to visit us now when you haven't shown your face here for years?' He paused. 'I bet it's got something to do with Aspect's arrest. You've probably also heard about the incidents in New York that the owner of this auction house must have been involved with. The things we know about them prove that you're quite wrong about your crimes of chance.'

'What do you know?'

'A few months ago, a colleague happened to hear about this puzzle the police in New York have been trying to crack for years. And you can imagine how dumbfounded he was when it turned out there was only one difference between the New York conundrum and our Soho mystery: over there, the district is spelt SoHo. The same break-ins with no recognizable system behind them. Rugs or jewels, door or window, upper or lower floors – all with the same frequency. We could hardly believe it. So we asked our colleagues in Manhattan for their data and compared them with ours.'

'And?'

'The details they sent us corresponded to ours. If a rug disappeared in Soho in London, a rug was pinched over there in SoHo, New York. There isn't one night we know of that the burglars stole something different. Not a single one! It's exactly the same for the floor they visited and their way into the building. Ground floor here, ground floor over there. Break-in through the door over here, break-in through the door over there. As if both had happened in the same place! Do you still believe in coincidences?'

'I don't know,' said Glock hesitantly. Then he asked, 'Why are you so sure the two gangs don't arrange things in advance?'

'I think that's out of the question. As you know, we

monitor all transatlantic data traffic, listen in on every word that anyone, anywhere and at any time says on the telephone, record every bit of data on every computer in the world. And the Americans don't do a bad job either. If the gangs were coordinating their raids – and for years, at that – they would have left traces on our servers. But we haven't found any.'

'So how do you explain that the same thing happens at the same time, 3,500 miles apart?'

'Simple. The burglars are following a plan that someone cooked up for them.'

'Aspect?'

'Of course. He runs auction houses in both Sohos. And it's in those two branches that the stolen rugs and jewels turn up.'

For a few minutes it was quiet. Glock avoided Stone's gaze and stared out into the fog. At one point he leapt up, briefly paced around the room, then sat down again. Finally, Glock broke the silence. 'May I have a look at your data?'

'As long as it's just between ourselves.' Stone opened his laptop, launched a programme and handed over the computer. Glock typed in a few commands.

'I see that you've recorded just one detail for every break-in. You either know whether a rug has gone, whether they broke down a door or whether the theft occurred on the ground floor. But you never know all three.'

'Yes, those are our new databases. We have the criminal psychologists to thank for that. They discovered that witnesses say more and more contradictory stuff the longer you let them talk. The data quality is best when each person states just one single fact. Now witnesses can only answer a single yes-or-no question, selected at random by the computer. For example: Were jewels stolen? And to make sure no investigator is tempted to ask more questions, they've set up all forms and databases in such a way that you can only enter that one fact for every event.'

'I've heard about that.'

'Fewer details are preferable to incorrect ones. They call it Heisenberg criminology, after the head of our scientific division. The prosecutors and courts were delighted when he pushed that through. Suddenly everything Scotland Yard says is correct!'

'But you have to make do with much less information than you used to.'

'Well, the criminals are to be convicted by collating data from as many similar cases as possible. Don't ask me what I think about it. At least it's going down well with the media, and the chief constable loves that. They're investigating according to the Heisenberg principle in the States as well now – always just one fact per case. We have to put up with it. For some nights we happen to know the same fact from London and New York.'

'Then you get your perfect match.'

'Exactly. If they got in through the window here on a particular night, then they did over there as well. If they burgled a ground-floor flat, then it happened that same night on the same floor in the US. For some nights we get different facts, so we know what they took here and how they entered the property over there, for example.'

'Let's forget New York for a moment and concentrate on London. You haven't found a pattern that lets you make sense of the break-ins here?'

'Unfortunately not. I've told you that already,' Stone replied impatiently. 'And, as you also know, it's not just us. Our colleagues in New York have no better idea of what's going on in Manhattan.'

'I know, I know,' Glock said. 'Isn't that strange? You know a lot about the connections between the burglaries in London and New York. But when you look at each series of burglaries on its own you're completely in the dark. You remind me of mechanics who claim to know everything about your car, but then when you ask them about the parts the car is made of they just shrug their shoulders.'

'I'm afraid that's how it is.'

'Then we'll just have to make do with the

information we have. So you know just one single fact about each burglary. Either you know whether they got in through the door, yes or no. Or you know whether they were on the ground floor, yes or no. Or you know whether they took a rug, yes or no. That's three yes-or-no questions. But for each burglary you only know the answer to one of these questions.'

'That's right.'

'Well then. The best thing to do is draw a chart.' Glock reached for a pencil, then presented this sketch to Stone:

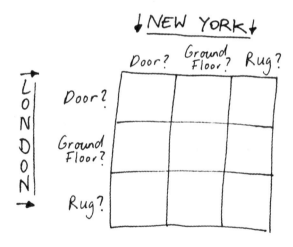

'Looks like a sudoku grid,' said Stone.

'It works a bit like that too. I start by choosing a question for London. And then I ask a question about the same night in New York. For example: did they pick the lock on a door over here? And was a rug taken on the same night over there? We have a box here for each possible pair of questions. OK?'

'Sure, why not. But what are you driving at?'

'I want to see how much information there is in your databases. We'll look at both now. We're going to search for all the nights on which our investigators mention a door lock being forced open and our friends across the pond say that a rug disappeared. As you can see, there are nine pairs of questions.'

'Right. And we don't need to ask any more of those questions, because if the answer to the "door" question was "no", we'll know they went in through the window.'

'For how many nights do you think we'll find the same answer in both databases?'

'You mean yes and yes or no and no? If we go through all nine sets of questions?'

'Yes, that's what I mean,' Glock said.

'That's obvious,' Stone quickly replied. 'For half the nights. We know that the thieves break open doors just as often as windows, get up to their tricks on the ground floor just as often as on the first floor and steal rugs just as often as jewels. The probability for each of these results is 50:50. So if you ask any of these questions, that will be the correct probability. Hand me the laptop, please.'

Stone entered a few commands. Thirty seconds

later a triumphant smile flitted across his face. 'Look,' he said. 'There it is. I asked the database to evaluate all possible pairs of questions. In 50 per cent of all nights you get the same answer for London and New York.'

'Yes, I imagined that would be the case,' said Glock. He thought for a moment, then said quietly but very firmly, 'Aspect will be released.'

'Sorry?'

'I mean there's no plan.'

Stone was speechless. 'What makes you say that?'

'Elementary, my dear Stone. You just have to count. Let's assume you were right and the gangsters were working to a plan. For example, the plan could say that on the first of April they would break in through a door on the ground floor in order to steal a rug. Now I can ask any of the following three questions for the first of April: did they enter through the door? Were they on the ground floor? Did they take a rug?'

'Of course,' said Stone. 'But I don't see what you're getting at.'

'I want to show you that your supposition that they're working to a plan leads to a contradiction. Because no matter which of the three questions I choose for London and which for New York, you have to answer yes every time. They entered by the ground-floor door both here and over there and stole a rug. So if that was the plan, then the probability that I will get two same answers is 100 per cent. Just have a look at it in the chart:

'I've written the answer for London on the left in each box, and for New York on the right. Of course they're both the same everywhere. No matter which questions I choose, I will get it right 100 per cent of the time. But just now you told me that according to your data the probability is only 50 per cent. Well, there you have it. Your assumption was wrong. There cannot be a plan.'

'What you're saying is only true for that particular case. It comes from the fact that the plan they're following always results in the answer yes, so of course they are all the same. But what if they'd been following a different plan? They could, indeed, have entered a ground-floor apartment to get jewels in both cities. So if you then asked me the question about the door and then the one about the rug from your catalogue, I would answer "yes" once and "no" once. So you don't always get the same answer from me.'

'True. But whatever the plan is, the probability that you will get yes/yes or no/no answers is always greater than 50 per cent. Because what options do I have of asking questions so that you will give me two different answers? Let's just have a look at that:

'Do you see? For four of the nine possible question combinations you have to give me different answers, for five you give me the same answer to both questions.

Therefore the probability of getting two identical answers is five out of nine, around 55 per cent, if the burglars were following a plan. Your result is once again greater than the 50 per cent your data suggested. So again, your assumption is wrong. And in all other imaginable cases you'll also end up with five out of nine. Whichever way you look at it, your assumption is wrong. There can't be a plan.'

Stone fell silent. Glock wrote a text message to Alice Aspect. Then he tried to look out of the window. The fog was so thick that all he could see was an expanse of grey.

'I can't argue with that,' Stone said after a long pause, 'but I don't understand it. If there really is no plan at all and if the gangsters aren't in cahoots with each other, how come that night after night the same thing happens in London and New York?'

'No one understands that,' Glock replied. 'The only way I can explain it is that there is some kind of inner connection between the two gangs. If something happens on one side of the Atlantic, then the same thing happens at the same moment on the other side.'

'That's what you're telling me? You, John Glock, the only physicist Scotland Yard ever hired straight out of university? You used to spend half the night lecturing me on Einstein: causes that set off an effect at the same moment somewhere else are impossible; even light, the fastest signal, needs time to make the journey. You've forgotten all that?'

'There has to be some reason for the mirroring of the burglaries. Or do you think that these crimes only exist as entries in your database?'

'Rubbish,' Stone replied. 'The jewels and the rugs are gone.'

'OK then. So the crimes are somehow so closely connected that the crooks don't need to exchange signals or have a common plan in order to coordinate the burglaries.'

'Telepathy? But that's ridiculous, John!'

'Not telepathy,' Glock said, crossing his arms. 'Nor a contravention of the theory of relativity. There's not the least bit of evidence to suggest that these criminals exchange messages. They just happen to do the same thing in two different places. What would be the simplest explanation for that? From the burglars' point of view, Soho and SoHo are the same place. The ocean between the two makes no difference to them. Of course, we find it more or less impossible to imagine such circumstances. But does that mean that no world exists outside of time and space? You know that for quite some time I've believed that all the places in the world are really just one place. And now we have some evidence for that! At any rate, you'll have to release Aspect.'

The events that Glock and Stone are trying to explain sound far-fetched, but they correspond to physical reality. We owe this insight to the French physicist Alain Aspect. In a famous experiment from 1982 he showed that there can indeed be some inner connection between events in different places that cannot be explained by prior arrangements or an exchange of signals. Because these events occur randomly, a secret plan or hidden transmission of information can be ruled out. And yet the same things do happen at the same time in places that can be as far apart as London and New York. Aspect's experiment is therefore a fundamental challenge to our concept of time and space.

The experiment is one of the most dramatic examples of the fact that reality does not conform to accepted logic. What happens in the course of it follows the laws of quantum mechanics, which determine the basic behaviour of energy and matter. In his experiment Aspect examined pairs of photons, the elementary particles of light. He constructed a sophisticated arrangement that would make it possible for criminals to do the same thing at the

same time in different cities without having to plan in advance or exchange signals. The information about which action is to be carried out is encoded in the photons.

Photons form a pair, a kind of community of destiny, when they are emitted simultaneously from the same atom. Arranging for that to happen is no problem. The two photons then separate from each other, yet remain linked in some mysterious way. In Aspect's original experiment they travelled a distance of 6m (20 feet); in a later experiment a photon went from Tenerife to La Palma, both in the Canary Islands and almost 150km (90 miles) apart. Austrian and Chinese scientists are currently running experiments to send one photon from a satellite to Vienna and its partner to China.

Once the photons arrive at their destination their condition is measured. Of course, elementary particles and their characteristics can be neither seen nor photographed – another thing they have in common with cunning crooks. Like Stone's method of investigation, the rules of quantum mechanics only allow yes-or-no questions. That is the decisive factor. Gradual transitions or in-between states are unknown to nature. We only experience them because we don't look at things closely enough – just like a printed black and-white picture has shades of grey from a distance, when in reality all it consists of is black and white dots. In the same way there is only yes or no, black or white, on the level of atoms and elementary particles. Nature is digital.

Quantum mechanics is therefore about the information nature transmits to the observer. For example, one could ask whether, at the point of measurement, a photon has already reached a particular place – yes or no? Another good question would be whether the particle is moving forward with a particular momentum. In his experiment Aspect asked about the intrinsic angular momentum, the spin, of the photons. The spin is a kind of inner direction of rotation of particles, an axis. For example, you can check whether this axis, the spin, is pointing upwards.

So Aspect kept on sending pairs of photons off on their way and asking whether the spin was pointing upwards. He made two observations: firstly that the answers came in random order. There were exactly the same number of 'yes' and 'no' replies but there was no detectable pattern in the sequence – as if the photons were tossing a coin. And secondly, the answers of both photons in a pair corresponded to each other. Once you had measured the answer of the first photon, you also knew the answer of the second.

A gang of thieves active in two cities could base their decisions on this information. If, for example, the result of the measurement in one city is 'yes', the crooks head out to steal jewels. The measurement at the same moment in the other city would then give the corresponding result. So those criminals know what their colleagues have done and can act accordingly – therefore jewels disappear in both places. The thieves neither have to schedule their activities in advance, nor send each other messages as they set about their task. In this way their robberies are perfectly coordinated and completely unpredictable.

Like so many of the great discoveries in physics of the last century this one also goes back to Albert Einstein. However, the astonishing realisation that events can be dependent on each other without an exchange of signals is one that Einstein came to against his will. In fact, his intention had been to show that precisely this did not work.

His aim was to undermine quantum mechanics, though he himself had prepared the way for that theory by realizing that light was a stream of photons. With the help of quantum mechanics people could understand for the first time what went on in the world of the tiniest units, and providing the impulse to that was Einstein's most significant achievement after the theory of relativity. But throughout his life Einstein was at odds with his brainchild. He once wrote, 'An inner voice tells me that this is not the true Jacob,' in other words, not the genuine article. In order to prove the inadequacy of quantum mechanics, in 1935 he devised the experiment that Alain Aspect was to carry out almost fifty years later.

Einstein didn't like the fact that according to quantum mechanics there are chance events, that is to say events that are fundamentally impossible to predict. Because quantum mechanics can only predict probabilities, for example as follows: if you set up an experiment in such a way that the results are answers to a yes-or-no question, the answer in 60 per cent of the cases will be 'yes', and in 40 per cent of the cases 'no'.

This means that nature is playing roulette and Einstein refused to accept that. He was convinced that the world is in principle comprehensible and calculable. 'You believe in a God who plays dice and I believe in complete law and order,' he once thundered at his friend and colleague Max Born.

Many physicists in the first half of the last century were absolutely devastated to discover that nature is indeterminate. But only Einstein saw one consequence of quantum mechanics that is even more disconcerting: the theory demands that once two particles are interacting they will remain forever linked. In Aspect's experiment the paired photons are linked together because they have a common origin. However, the linkage can also come about if the two particles influence each other later in their life history, because at some point a force is established between them.

Such linkage is called entanglement. Its essence lies in indestructible information about the history of the particles. This information is not even lost when the two particles move away from each other, no matter the distance.

That is the point from which Einstein's deliberations started out. If you measure the state of a photon in London, you get a random result. But an entangled photon in New York must give the opposite result at the same moment; after all the two particles are entangled, inextricably linked with each other. That is exactly what Aspect was able to observe. And that is exactly where Einstein saw the contradiction. For if the result in London was random, how could the New York particle have learnt about the random result on the other side of the Atlantic so quickly that it could deliver the same result? Such a thing was impossible, Einstein declared. According to the theory of relativity, signals spread at the speed of light at most, which is why they take time to get from here to there. But the entanglement effect appeared immediately – quantum mechanics demanded 'spooky action at a distance', Einstein mocked. Therefore quantum mechanics had to be wrong, he went on, the supposed random results were not random at all. There was some hidden plan behind the remarkable events. Einstein had realized that nature defies our logic. He refused to accept it.

How do you find a hidden plan? How do you determine whether it exists at all? For decades these questions seemed to defy solution. But in 1964 the Irish particle physicist John Bell had an idea that matched Einstein in its astuteness: a hidden plan would leave traces. And all one had to do to find these traces was count. If there was such a plan, then it must of necessity change the frequencies with which certain events appear. Dealing with quantum mechanics usually means grappling with complicated mathematics. However, John Bell revealed the disturbing core of quantum mechanics with primary school calculations.

John Glock in the crime story employed this principle by counting frequencies – for example how often he could get the same answer to two yes-or-no questions about burglaries in different places. He showed the dumbfounded Stone with his calculations that such pairs occurred slightly more frequently if the event had been arranged beforehand than if it was based on pairs of coincidences.

In his original experiment Alain Aspect was asking about the direction of the spin of his photons. He tested how often the results of the measurement of the two paired photons agreed if he asked different questions, that is to say if he turned the devices measuring the spin towards each other. Then he counted how often the data agreed. From the result of that he could conclude that there was no hidden plan. Had the paired photons been following such a plan, the hit rate would have been different.

And now that later experiments have cleared up the last remaining doubts, it is certain that there cannot be a hidden plan determining the fate of each and every particle. They actually are coincidences that come into effect simultaneously in places far apart. Einstein was wrong.

The experiments with entangled particles were also known as 'beaming', after the famous transporter on the Starship Enterprise that could instantly transfer Captain Kirk and his crew to unknown planets. Today entangled photons can be dispatched over hundreds, soon over thousands of kilometres. By now these connections are even being used to transmit messages securely, without the risk of being intercepted.

And it is no longer just elementary particles that can be beamed. Physicists in Oxford even managed to entangle two diamonds. The precious stones were almost the size of a fingernail. If one of the stones was asked about its state, both diamonds, like magic crystal balls, immediately gave the same unpredictable answer.

The difficulty in such experiments lies not in bringing about the entanglement but in preventing it. The art is to isolate the entangled objects so effectively from their surroundings that they don't transmit any information to other items. Otherwise the entanglement would be diluted, so to speak, and, like a fragrance dispersed by the wind, its presence could no longer be ascertained. That is why we don't perceive the entanglement in our everyday life – it's ubiquitous.

Einstein's objection however remains uncontradicted: entangled objects behave as if there were no space between them. If one partner gives a random answer when measured, the other gives the corresponding answer at the same moment, no matter whether it is just a millimetre or 100km away. How is that possible? Could there be internal connections that disregard the framework of time and space? Are near and far simply aids to orientation for us but of no significance at a deeper level of reality?

It is difficult for us to accept that entanglement is not something 'spooky' but real. We can imagine a world governed by random events, but not one without space. Our everyday experience compels us to imagine space as a kind of box into which everything is poured. It is in space that we see, hear and feel everything that can be observed. A world without space, it seems to us, can only be a spirit world. In such a world we feel like strangers.

We perceive space as an anchor for our experience, just like time and consciousness. Therefore we are tempted to see these three as beyond explanation. But the passing of time, as we will see in chapter 8, is just the decay of order. And there is a lot to be said for the idea that consciousness consists of the collaboration of billions of neurons. Why then should space of all things be fundamental?

It is possible that what we experience as space is just a very crude representation of the relationships between things. Left and right, above and below, in front and behind could be derived from the fact that objects are entangled with each other: in that case closeness would be, as in human relationships, just another word for a particularly strong connection. Perhaps space is not a box but a net, held up by everything around us. In that case Glock would be right and all the places in the world are in reality just one place.

Weeks later, when Glock was long back at his desk, sitting motionless and staring out at the London drizzle, full of hope that he would soon have completed his manuscript, a man in a dark suit appeared at his door. The stranger sketched a bow and, without a word, handed him an envelope of Japanese paper.

Glock nodded, put the envelope down and set off to meet Alice in Central Park in half an hour.

6

Is the World Real?

A hammer hits a thumb. But the hammer,
like all matter, consists of emptiness.
How can nothingness hurt like that?
And then   does the nothingness exist at all?

When I pronounce the word Nothing
I make something no nonbeing can hold.
WISŁAWA SZYMBORSKA

In one of the most successful films ever made, people become victims of a complete delusion. Houses rise up into the sky, the streets are full of people who feel like men and women and also smell like them. People make friends or squabble with their colleagues and partners – they live in a world just like ours. Only an elect few know the truth: the houses and streets are not even backdrops, their fellow human beings not even dolls. All of that, in fact everything the victims experience, exists only in the so-called Matrix. The Matrix is a gigantic computer programme that simulates a virtual reality, through which all-powerful machines have destroyed the free will of *Homo sapiens* and enslaved humanity.

The fact that *The Matrix* and its two sequels grossed way over a billion dollars surely cannot simply be explained by the unease of people who feel more and more dependent on their computers. I found the films unsettling as well, yet I certainly don't believe in a conspiracy of computers against humanity. Even if some contemporary philosophers try to demonstrate that, with a probability close to certainty but naturally without our noticing, we are all living in a simulation with which a 'posthuman civilisation' is duping us – well, I'm not convinced.

But I can no more believe that reality is the way it appears to us. Naturally I'm not the first to be bothered by such scepticism. That the world might be nothing but an illusion is an age-old idea. It is discussed in the earliest writings of Indian philosophy from almost three thousand years ago. In the Vedas, what Hollywood called 'the Matrix' is given the name 'Maya' – a deception that is difficult to see through which conceals a quite different reality, or none at all. Later the Buddhists contended that the void alone was real. The things and bodies around us had no substance at all: the significance we ascribed to them existed solely inside our heads; all perception was just a dream from which most people never woke. Thinkers in ancient Greece expressed similar ideas. These philosophers in the East and West were also in agreement that the most rewarding goal in human life was to free oneself from that illusion.

We don't lose sleep over these basic doubts about reality. We simply hold on to what we can touch, the matter that everything around us clearly consists of. And even if we have our doubts about whether the world actually is the way it appears to us, such lines of thought cannot withstand our sense perceptions for long. After all, we can see, feel and smell that the world outside is there. Thus, after a few moments, experience overcomes scepticism. The New York philosopher Sidney Morgenbesser was once asked why there was something rather than nothing. 'I will tell you the reason,' he replied. 'If there were nothing you'd still be complaining.'

But the doubts calmed by the drug of obviousness continue to affect us deep inside. When a film such as *The Matrix* stirs them, they resurface. When you see the world through the eyes of the hero, Neo, and follow him as he exposes his experiences as a delusion and eventually penetrates the code of the Matrix, we wonder whether that could happen to us as well. Perhaps what we think of as reality is just an illusion after all. It doesn't have to be computers that are hoodwinking us. The delusion could have quite different origins. The scenes in the film haunted me because they made me think of physics. In order to dispel our doubts about the world, we rely on matter: the table we touch is solid and undoubtedly real. But if we look at matter more closely, it seems to disintegrate and lose every single characteristic we expect it to have.

So is matter itself an illusion? But then who or what can guarantee that the world is more than just imagination? Clearly it's worth our while to look at that question more closely.

The chair I'm sitting on bears my body. I can feel the pressure of my hand on the table. Objects have volume and mass, they're firm, they resist us. That seems to be the way things around us are constituted. Even the air has to be pushed aside when I move around in it. And anyone who's just hit their hand with a hammer and sees their thumb turn blue has a hard time believing that the hammer only exists in their imagination.

The hammer is matter. It's solid. Like everything around us on Earth, it consists of atoms. It is impossible to doubt the reality of atoms. For some years now we've even been able to see atoms, the scanning tunnel microscope shows them to us as cloudy shapes. In the iron crystals of the hammer the atoms stand lined up in rows like soldiers.

But what are the atoms? Real clouds consist of something: water vapour, raindrops, ice. However, there's almost nothing at all in the cloudy shells of atoms. There are just a few tiny electrons whizzing around – in iron, there are 26 of them. Otherwise the atom clouds are empty. At the centre is an atomic nucleus. But the nuclei are incredibly tiny, while the clouds are huge. If you were to blow up an atomic nucleus to the size of a midge, then the shell would have to assume the dimensions of a concert hall.

In this huge edifice the midge would be lost. What then is there between the midge and the walls of the concert hall, that is to say in the atomic shell? What is there inside the hammer? Empty space. Nothing. How can nothing be so painful?

The comparison of the atom with an empty concert hall comes from Ernest Rutherford. He was the first to see that apparently solid objects are, in reality, empty. He described this discovery as 'the most unbelievable experience of my life'. Rutherford was one of the most productive experimental physicists of all time. Born in 1871 as the fourth child of a farmer in the far south of New Zealand, he came to England by steamship after completing his degree and started to unravel the secrets of radioactivity at Cambridge University. Astonishingly early, and long before his colleagues, he realized the consequences of what he was doing. As early as 1903 he described the forces that can be released from an atomic nucleus. Whoever succeeded in setting off a chain reaction, he wrote, could 'make this old world vanish in smoke'.

Together with his German colleague, Hans Geiger, who later invented the Geiger counter to detect radioactivity, he carried out the key experiment in 1911. The two physicists bombarded gold foil with particles that were only later found to be helium nuclei. Almost all the particles flew straight through the gold as if it wasn't there. The experiment proceeded as if Geiger and Rutherford had been shooting at empty space. Just one in 20,000 projectiles rebounded, but that with enormous force. From this more than surprising result Rutherford deduced three things: firstly that any particles that came back had hit the atomic nuclei of the gold. Secondly that almost the whole of the mass of gold was gathered in the tiny nucleus, while the shell with its electrons gave the gold its volume but hardly made any contribution to its mass – the midge was a thousand times heavier than the concert hall it sat in. Thirdly the huge shell, the cloud round every atom, consisted literally of nothing. The most everyday objects are as empty as outer space between the galaxies.

However, Rutherford thought that at least the nucleus of the atom was solid. There must, he assumed, be something firm to hurl back the projectiles. But even in that Rutherford was mistaken. The atomic nucleus is also empty. Today Rutherford's intellectual heirs, the particle physicists, bombard such nuclei in order to tear them apart. The projectiles pass through gigantic accelerators in order to reach the necessary speed. The equipment of the European Centre for Nuclear Research (CERN), for example, occupies a 26-km (16-mile) tunnel ring near Geneva and is, therefore, the biggest machine ever constructed. I have visited CERN several times; in no other edifice, not even standing beside the pyramids, have I ever felt so tiny as in those caverns deep in the rock, where particle detectors rise up to the height of skyscrapers.

These detectors proved among other things that the supposedly solid atomic nucleus is for its part composed of elementary particles, the quarks. Thus the atomic nucleus is not a solid entity either but also a cloud. In that cloud the quarks are whizzing round, with emptiness between them.

In effect, the whole atom therefore consists of emptiness. For although the electrons in the shell and the quarks in the cloud undoubtedly exist, they don't take up any space. When measured in the accelerator they don't show any dimensions at all. Anyone who has trouble imagining a thing without size is in very good company. Physicists constantly kept coming across such paradoxes when they started to explore the atom in the first half of the 20th century. The apparent contradictions come from the fact that we automatically try to apply our everyday way of seeing things to a reality that is of a quite different nature. We have transferred our familiar experience, that things are solid and need room, to the atom. We're like the innkeeper who is said to have asked what to feed the horses when Bertha Benz (the wife of Karl Benz, designer of the world's first practical automobile) parked her husband's car outside his inn during her first long-distance drive.

But matter does not correspond at all to the way we experience it. Its component parts have no firm contours. Elementary particles are not little spheres but conditions of energy, a sudden flaring up in empty space. 'It is probably as meaningless to discuss how much room an electron takes up as it is to discuss how much room a fear, an anxiety or an uncertainty takes up,' the English physicist James Jeans wrote. Perhaps we can imagine elementary particles as points someone is dotting in space with the infinitely fine tip of a pencil. The points move, they come and go. The things that seem to us so corporeal are in reality just a skeleton of dancing marks in the void – like the little numbers on dot-to-dot pictures that children join up to make a drawing.

Other particles are journeying hither and thither between the dancing corner points. It is only through this that a connection is established between the components of the atom. The nucleus and the electrons in the atomic shell exchange photons – the elementary particles of light. Photons are pure energy. They transmit the electromagnetic force.

How does that happen? Richard Feynman, who gave us cause to reflect on the beauty of the rose in the first chapter, came up with a vivid image for the interplay between particles: when an electron emits energy, a photon is born. This photon receives the energy that the electron now lacks. It can fly to an atomic nucleus and transmit the energy to it. When the nucleus has absorbed the energy, the photon has disappeared. But that nucleus can now emit energy and spit out a photon that travels to the electron. Thus the nucleus and the electrons are linked by the exchange of photons. This image can be translated into formulas that physicists use daily in their calculations.

Thus an order arises from the exchange of photons, just as an order arises on the football pitch once the ball comes into play. The players of both teams automatically take up positions where they are best placed to pass or receive the ball. The photons have a similar effect to the ball on the field of play: they ensure the correct spacing, determine the scale of the whole event. They also hold the atoms together and make clouds, drops and crystals grow through the electromagnetic force. The photons create order.

That is why we have the illusion that we live in a world of solid things. The void takes on form because particles, which have neither shape nor extent of their own, enter into a relationship with each other. A bond between spooky objects makes us perceive the hammer and thumb as spatial entities. What is more, it ensures that the blow of the hammer has a painful effect. The bond of the particles is strong in both the iron and the flesh of the thumb. Therefore the void of the hammer cannot simply pass through the void of the thumb.

However, anyone who picks up a hammer doesn't just feel a form but also a mass. Therefore we have to exert force to accelerate the hammer in the direction of the nail. The mass of the iron is responsible for the pain if the blow misses.

So where does this mass come from? The hammer consists of a cluster of particles without volume. How can mass be stuffed into such bodiless entities? No one would think of putting light or pain on a pair of scales. But the particles have entered into relationships, and every bond contains energy. The energy of the bond between electrons and the nucleus, for example, is contained in the photons that hold the two together. And, as Einstein taught us, there is a mass corresponding to all energy. Mass accrues to spooks that unite.

Far and away the strongest bonds are at work in the atomic nucleus. There, gluons – 'sticky particles' – shuttle between the quarks, just as photons mediate between the nucleus and the electrons. When an atom bomb explodes, part of the immense energy of the gluons is unleashed. We feel this same atomic energy as mass when we pick up an object and in the power with which a hammer lands on its goal. But this energy, which we experience as mass, is not simply there. It only arises when particles enter into a relationship with each other.

The particles do in fact have a mass of their own. However, it is very small compared with the mass that arises from their bond. If you could put each of the quarks and electrons the hammer consists of on a pair of scales, their combined weight would only add up to that of a sheet of paper. But even if we can hardly feel the mass of the particles in the hammer, we mustn't simply disregard them, for those couple of grams are decisive: it is only because quarks and electrons have a mass of their own that they can enter into a bond at all. Their mass makes the particles 'heavy' and resistant to acceleration. Massless quarks and electrons would zoom past each other at the speed of light and never form atoms.

But how do particles that have no extent come to get their mass? This question occupied physicists for decades. Only in 2012 did they manage to solve the mystery in the Geneva accelerator. In preparation, the scientists had once again upgraded their machine, so that it was now capable of generating a simply unimaginable concentration of energy: the device bundles enough energy into less than a millionth of a gram of protons to melt a ton of copper. Powered by this, the particles zoom along through almost complete emptiness at 99.999999 per cent of the speed of light. The gas pressure in the ultra-high-vacuum pipes of the circular collider tunnel is ten times lower than on the Moon.

The particle physicists discovered that absolute emptiness did not in fact reign in the vacuum. The void is filled by something that the physicists called the Higgs field, or the Higgs for short, after a Scottish colleague. The Higgs behaves differently from everything we know. It is completely permeable to light, without form, not directly detectable and yet omnipresent.

It is best to compare the Higgs with an immense snowfield. Anyone walking across a wintry landscape in dull light will no longer perceive the snow – all they see is a shapeless expanse of white. Similarly we can neither perceive nor measure the Higgs field because it surrounds us evenly and completely. But our walker will sense that there is something offering resistance to their every step, as they sink into the snow. In exactly the same way the Higgs field hampers the movement of the particles. It makes them 'heavy'.

When the physicists in Geneva announced their results to the world, the news reports mostly talked about the discovery of the 'Higgs particle' rather than a field. Higgs particles are something like the flakes in a snowfield: a tiny structure in a uniform expanse, which – in contrast to the field itself – can be demonstrated. And that is what they managed to do at CERN.

Was it worth spending more than ten billion Euros on this? The discovery of the Higgs particle from characteristic traces of its decay showed that the theoreticians were right. Now it had been proved that the otherwise invisible Higgs field did actually exist. Thus, we also know where particles of matter get their mass, even though they themselves have none of their own. As soon as a quark or an electron wants to travel, it is as if it were tramping through snow: the field offers it resistance, the particle becomes 'heavy'. The resistance suggests to us that it has mass, but in reality what we are experiencing is the viscosity of the void.

In addition, the Higgs explains why not all particles are equally heavy: it hinders some more, some less. A skier will glide quickly across the landscape, while our booted walker will sink up to their knees. A bird isn't bothered by the snowfield at all, it simply flies over it. Electrons are skiers, quarks walkers, photons birds – the first proceed quickly in the Higgs field, the second slowly and the third completely unhindered. Thus electrons have a small and quarks a large mass, while photons have no mass at all. It is only because that is the way things are that there can be stable matter. But the Higgs itself is not matter but a product of the vacuum.

This remarkable field acquired its characteristics when the universe expanded and cooled down after the Big Bang. Before that, in the unimaginably hot cosmos, it was still so flexible that it didn't offer resistance to any movement. At that time, all the particles were flying around at the speed of light; there was neither mass nor were there bonds. Every particle was a loner. But just as water freezes to snow or a block of ice when the temperature falls, the Higgs acquired its viscosity. Since then, the elementary particles differ in the extent to which their movement is hindered by the Higgs. We experience this hindrance by the vacuum as mass. It is through it alone that forms take shape in the cosmos. What we have learnt through the expensive experiments in Geneva is that we owe our existence to a whim of the void.

So is the world real? It is at least constituted completely differently from the way we experience it. We are actually living in a matrix. But we're not the victims of an illusion created by an all-powerful computer. Instead we are of necessity ourselves making do with illusions to find our way around a bewilderingly complex reality.

Gottfried Wilhelm Leibniz would have been delighted by these discoveries. 'Why is there something rather than nothing?' was the question he asked himself. Leibniz was an ingenious philosopher, mathematician and inventor, who around 1700 served as librarian to the Duke of Hanover and was one of the most profound minds Germany has ever produced. He commanded such admiration that a kind of biscuit was named after him. Leibniz was rarely at a loss for answers. He invented calculating machines, thought up philosophical systems and, in competition with his contemporary Isaac Newton, discovered the fundamentals of modern physics.

The question as to why the world exists at all 'is the first question one is justified in asking,' he wrote. And he was left speechless in the face of this question of all questions. 'For nothing is simpler and easier than something.'

However Leibniz could only conceive of the void as complete emptiness. He failed to see that it is pointless to reflect on such an absence of everything. For in that case you have to tacitly accept that something exists that is absent. The very definition, therefore, contains an internal contradiction. An all-embracing void, as Leibniz conceived it, would be like unicorns guarding a cosmos without any unicorns in it.

Matter is quite different from the way we imagined it. The same is true of the void. It is not a state of absolute emptiness but one of formlessness. The void is a stage without a play, a space where anything can arise. Everything we experience corresponds to a play on that stage, to an order in the void, that flares up and then vanishes. We call it hammer and thumb, earth and heaven, man and woman.

7

'Who ordered that?'

We live in a shadow world. No matter where we look, there is twenty times more than appears to us. More of what? We have no idea. But without dark energy, without dark matter we couldn't exist.

I like looking at old maps. The most beautiful ones come from the days when people sailed the oceans but the continents were still full of secrets. I then imagine myself, one or two generations after Christopher Columbus, setting off with one of those documents to explore the world. On the coast of Central America, for example, a map from 1606 would have served as a good guide. The outlines of the coast, the estuaries and the settlements that were just arising are in fact marked with such fine strokes that they are hardly less precise than modern maps. But as soon as you go off into the interior, you are faced with emptiness. At best, the adventurer can follow the course of a few rivers winding their way across the blank land mass. To the left and right of them and at their sources, the cartographer has drawn little molehills, presumably to indicate that he suspects there are mountains there. Back then, no European eye had seen them. Then finally, beyond the watershed, in the middle of the continent, we see half-naked people and the word *Canibales*. I'm both fascinated and moved by the way the draughtsmen combined precision and imagination in one image because they knew that beyond the lines of the coast, which had only just been surveyed, there had to be so much that was unknown that even centuries later explorers would return home with surprises. Suddenly, the world had become much bigger.

When I look up at the night sky, I have a sense of how mariners must have felt off the coasts of the new world. However impressed I am by the expanse across which my eyes roam, I'm as overwhelmed by the knowledge that much more, that we don't know, is hidden behind what is visible. For the explorers, the coast was just the boundary at which the land mass of America rises out of the ocean – and the night sky is just the shining outpost of a much bigger, invisible world.

For a few years now we have known a figure that every person of an inquiring mind must find disturbing: almost 85 per cent of the cosmos consists of a totally unknown substance. To put it another way: no matter where we look in the sky, there is at least five times more than we can actually see. More of what? We have no idea. The unknown something has so far eluded every investigation.

We live in a shadow world, among huge masses that do not reveal themselves and about which we know nothing. But the shadow world does make itself known to us by painting ghostly images on the night sky. Stars suddenly change their brightness, galaxies appear surrounded by rings, enlarged, distorted, mirrored two or four times over. Some of the phantoms recall smiling faces, others appear as crosses in space. One of the most beautiful of these shapes is the Einstein Cross in the Pegasus constellation, which wasn't discovered until 1985, even though amateur astronomers can detect it with their telescopes. It consists of four spots of light of the same brightness and the same shape; they mark out the arms of a Greek cross. As in a kaleidoscope, the four spots represent four times over the same very bright object which is more than eight billion light years away. In the middle of the cross there's the glitter of a galaxy. The strong, quadrupled source of light is, as its spectrum reveals, a quasar: there is a black hole concealed within it that sucks in matter and accelerates it so fast that it flares up brightly one last time.

In 1985 we were literally still in the dark about the true secret of such formations. It was thought they could only be explained using the theory of relativity: seen from Earth, the galaxy and the quasar were exactly aligned, and in such a way that the galaxy covered the quasar. But, as Albert Einstein taught us, large masses warp space, meaning that the light close to such masses no longer spreads in a straight line but describes a curve. This prediction had already been proved during the great solar eclipse of 1919: when, on the morning of 29 May, the stars appeared over South America, they seemed slightly out of place and the nearer they were to the darkened sun, the more they had changed their position. The mass of the sun had dented space, bending the stars' light rays. The *New York Times* ran the headline, 'Lights All Askew in the Heavens. Einstein Theory Triumphs', while reassuring its readers that, although the stars weren't where they seemed or were calculated to be, there was no need to worry.

And the galaxy in the middle of the Einstein Cross also warps space. However, since it contains well over a billion stars, it can produce a more complicated warp than the sun. Thus the light from the quasar can circumnavigate the galaxy by different ways and then appear quadrupled in the sky. That is why astronomers initially saw no reason to find the phenomenon surprising. They thought that spooky images such as the Einstein Cross were just another, though admittedly spectacular, confirmation of the general theory of relativity.

Einstein had even predicted that gravity can distort space into a lens through which distant galaxies appear enlarged, as if through a gigantic magnifying glass in the cosmos. In recent years such cosmic magnifying glasses have indeed been found. One of the most powerful of these gravitational lenses is a huge galactic cluster called Abell 1969 in the Virgo constellation, more than two billion miles away from Earth. The galactic cluster distorts space, bending rays of light to such an extent that far-away galaxies appear as circular lines and all objects beyond Abell 1689 as enlarged. Thus it was that Abell made it possible in 2008 for astronomers to discover some of the oldest and most distant objects in the cosmos. Through this natural telescope they saw a galaxy on the edge of the visible universe that arose a few hundred thousand years after the Big Bang.

But the real surprise was Abell 1689 itself. As it turned out, there were far from enough stars in the galactic cluster to set off such dramatic effects. And no other accumulations of mass that could distort cosmic space to that extent are anywhere to be found. But it is indisputable that the sky, seen through Abell 1689, seems enlarged and studded with haloes that could only come from a severe distortion of space. So what can have caused it?

There are two possibilities. Either the known laws of gravity are wrong – at great distances gravity has a stronger effect than was thought, Einstein and before him Newton were mistaken. A minority of

astrophysicists actually believe that. I, however, am not convinced. Under all other conditions the law of gravity has stood the test. Anyone who wants to abandon it should have a better alternative to offer. Nobody has found one.

And there is a much simpler possible explanation for the severe bending of light round Abell 1689: the law of gravity is correct, but Abell 1689 contains much more than we can see. The stars are just a small part of the agglomerated masses in the galactic cluster. This is what the majority of astrophysicists assume and it's what I believe as well. Accordingly Abell 1689 must for the most part consist of substances that do not emit any light, nor any other radiation – dark matter.

Its gravitation however, which bends the light, does reveal the presence of dark matter. From the strength of the cosmic magnifying glass, the form of the otherwise inexplicable light rings and the spooky images, we can work out where the invisible mass is concealed. Thus every galaxy in the Abell 1689 cluster is surrounded by a halo of dark matter, a kind of cloud, that weighs ten times, a hundred times, occasionally even a thousand times more than all the stars put together. And then, finally, the whole galactic cluster is surrounded by a halo of dark matter.

It is to this cloud that Abell 1689 owes its existence. Without it the galactic cluster would have long since disintegrated, or possibly not even have arisen at all. The galaxies that make up Abell 1689 are moving relative to each other so quickly that they experience huge centrifugal forces. They would long since have spread out in all directions of the cosmos if the cloud of dark matter were not holding them together with its gravity. Even the galaxy that makes the Einstein Cross in the night sky is nine-tenths dark matter.

And that is the way things are everywhere in the universe. It is solely because dark matter holds luminous matter together that there are any structures in space at all. Dark matter is cosmic cement. Every galaxy, ours included, is wrapped in it. In 2014 astronomers finally discovered the first dark galaxies – the Milky Way has a dark twin in the cosmos. It hangs out in the Coma Berenices constellation, and is a galaxy whose dimensions are similar to ours but which consists almost entirely of dark matter.

Dark matter doesn't just reach far out into the cosmos in the vicinity of the luminous galaxies, it also forms its own structures and fills intergalactic space. The luminous galaxies appear to be sitting on it like blobs of cream on a gigantic chocolate cake. Is everything we can see out in space merely decoration?

The cosmologists of the 21ˢᵗ century seem to me like explorers of previous ages who, after landing on a distant continent, are faced with a gigantic coastal mountain range and are now wondering what might be hidden behind it. They can't even see a pass because the mountains are completely covered in fog.

All we actually know about dark matter is that it doesn't shine or emit any other kind of radiation. Therefore it must be of a quite different nature from the world we know: there are electromagnetic forces at work in all visible things. Electromagnetism joins up atomic nuclei and electrons to form atoms, the atoms to form liquids and solids, and sees to it that there is resistance when various liquids or solids meet. And light is nothing but electromagnetic radiation. Dark matter, on the other hand, is not responsive to electromagnetic forces. Therefore it remains unnoticed; like ghosts, dark particles can pass unhindered through everything that is visible. According to some estimates billions of such particles are passing through our bodies every second without us knowing anything about it. For dark matter we're not really there at all.

And yet the shadow world of dark matter exercises cosmic influences, because with its mass it attracts other masses. Dark matter has an effect on us simply through its gravitation. In so doing it not only gives our Milky Way its familiar spiral structure but possibly exerts direct influence on earthly events. According to particle physicist Lisa Randall, it might even be responsible for the extinction of the dinosaurs. Randall, who is carrying out research at Harvard, wondered why large comets regularly hit the Earth. She believes she can see a pattern in the sequence of these catastrophes: roughly every 30 million years our planet is hit several times by other heavenly bodies. The cause, she says, is dark matter agglomerating in the median plane of the Milky Way. Because the Sun, like all stars, rotates around the centre of the Milky Way, it passes through that median plane roughly every 30 million years. That is when, Randall suspects, the dark matter gathered there sets comets on a collision course with Earth with its gravitation. This would explain, she goes on, why 66 million years ago a gigantic meteorite tore into what is now the Gulf of Mexico, its impact setting off a climatic catastrophe in which the dinosaurs perished.

At the moment, no one knows precisely how dark matter is spread over the Milky Way and whether it is in fact capable of diverting comets into the Earth. Therefore Randall's theory remains speculation. But even if it should turn out that dark matter was not directly responsible for the end of the dinosaurs and the rise of the mammals, it has determined our destiny. For without it there would have been no destiny whatsoever, neither for us nor for any life at all in the cosmos. It was only through gravitation coming from the dark masses that visible material, which spread evenly throughout the cosmos after the Big Bang, could gather in galaxies or stars could ignite. Without dark matter our solar system would not exist – we wouldn't be here. It is to this great unknown in the cosmos that we owe our existence.

Dark matter is one of the greatest challenges to our understanding of the world in the 21st century. Physicists have spared no effort finally to pin down these fateful particles, of which much of the universe consists. They have launched detectors to the international space stations and buried themselves in tunnels thousands of metres below the surface of the Earth, where they hope to be able to measure a tiny kickback when a dark particle makes a direct hit on the nucleus of the atom of a noble gas. Such a signal would be so weak that only kilometre-thick layers of rock could provide protection from interference. It is also impossible to shield the detectors with freshly mined lead, since it still gives off weak radiation. That is why Italian physicists tasked divers with salvaging metal from ancient Roman shipwrecks at the bottom of the Mediterranean Sea. To the horror of archaeologists, they carted lead from a ship that had once probably belonged to Emperor Nero off into their tunnel. Physicists have carried out almost a dozen of these tunnel experiments in the Apennines, and in abandoned mines in the United States, China and Canada. But despite a decade of effort not even a trace of a single dark particle has appeared.

Dark matter not only eludes measurement, it defies theoretical explanation, for it doesn't fit into the picture of the structure and origin of the world that physicists developed in the second half of the 20th century. The 'Standard Model of Cosmology' is what physicists somewhat diffidently call their version of the story of creation; the name is far too modest. Although this theory cannot cope with dark matter and therefore a large part of the cosmos, it is one of the greatest triumphs of science, for it describes how all matter known to us, and therefore the visible universe, developed out of the energy of the Big Bang. Basically the Standard Model tells a story, hidden away in mathematical formulas. This story is about how more than 13 billion years ago the first particles evolved out of pure energy; how the particles coalesced into atomic nuclei; how the nuclei captured electrons and thus became atoms; how the atoms gathered together in clouds of gas, stars and galaxies; and finally how the stars, acting as natural fusion reactors, bred the chemical elements, including oxygen and nitrogen, of which our bodies mainly consist.

Unlike a myth, a scientific creation story has to stand up to verifiable data. And the Standard Model passed all those tests with flying colours. It was confirmed by every astronomical measurement and by all the experiments of particle physics that were ever carried out. The Standard Model of Cosmology has a sibling, the Standard Model of Particle Physics. The latter doesn't tell a story but describes an order – the order of the construction kit from which the whole of the visible world is made up. This construction kit contains just a dozen different material particles. The most important ones are quarks, of which atomic nuclei consist, and electrons that circle the nuclei. Alongside those there are a few exotic relatives of the electrons and finally the Higgs, which provides all the particles with their mass. All visible things are made up of this handful of ingredients.

Four forces between the building blocks hold everything together. The first is called the 'strong' force. It binds the quarks within the atomic nucleus, which was why it used to be called the strong nuclear force. The second force is the 'weak force'. It causes radioactivity by making certain particles decay. These two forces only come into effect when two particles come very close to each other. The third is the electromagnetic force. It ensures that atomic nuclei and electrons combine to form atoms and is responsible for all electromagnetic radiation, for example micro- and radio waves. The fourth force is gravity. Like the electromagnetic force, gravity can also operate over very great distances.

According to the Standard Model there should be nothing more than the known particles and the four forces. Dark matter has no place in the world construction kit. 'Who ordered that?' the American Nobel Prize winner for Physics Isidor Rabi is said to have complained when an unexpected relation of the electrons turned up in the particle lab shortly before the Second World War. Today dark matter puts physicists in a similar but greater dilemma.

They have to live with the fact that more than 80 per cent of the material of the universe has turned out to be dark. What is even more mysterious is why the Standard Model, despite its huge gaps, has proved so successful. If you remember how little we know, you can only be amazed at how well we understand the cosmos. Or are we just enjoying the illusion of understanding?

The age of great discoveries is not over, it is just beginning. Anyone who, like me, believes in the existence of dark matter, accepts that this most widespread substance in the universe is totally unresearched. Anyone who, like a minority of astrophysicists, doesn't believe in it, is therefore compelled to dispute the laws of gravity as they are known today. In either case, we are still faced with uncharted territory. And whichever way you look at it, dark matter or strange gravity are far from being the most substantial unknown contents of the universe. What is also unexplored is the greatest part of the energy filling the cosmos. For all forms of energy we are familiar with are far exceeded by an unknown form. It is beyond doubt that this exotic energy does exist and determines the destiny of the universe. We know nothing about its nature.

But then, how are people supposed to be able to understand a cosmic phenomenon, the existence of which was only discovered a few years ago? In 1998 two groups of American, European and Australian physicists were measuring the speed of distant galaxies. That they are moving away from us has been known for decades; the flight of the galaxies was seen exclusively as a consequence of the Big Bang. The energy that was released at the beginning of time is still driving space apart – like leavened dough left to rise in a warm place. However, the Big Bang happened 13.8 billion years ago, so it was to be expected that the expansion of space would gradually abate. No pizza dough keeps on rising forever; once the yeast has used up the energy of the sugar in the mixing bowl, the whole lot collapses again. And also at work in the cosmos is gravity, which, in drawing all things together, counteracts the tendency to expansion. Therefore astrophysicists worked on the assumption that the expansion of the universe was at least slowing down. Some scientists even assumed that the phase of dilation is over and the universe is already collapsing. For a cosmos that is the way we have imagined it thus far would have to implode at some point due to the forces of attraction of visible and dark matter.

What the measurements revealed was all the more surprising: the universe is showing no signs of coming to rest or collapsing. On the contrary, its expansion is actually accelerating. Gravity, which draws all masses together, is being counteracted by something unknown.

'Dark energy' is what the American Michael Turner called this force that is making the universe grow faster and faster. He declared it the greatest mystery of all science, and one can hardly disagree with him. Dark energy makes the cosmos resemble the sweet porridge in one of the fairy tales of the Brothers Grimm, which keeps flowing out of the magic porridge pot and soon fills the whole kitchen, then covers the street, buries the town and doesn't stop rising. It seems inexhaustible.

This effect cannot be explained using the known forces of nature. All four forces known to physics – the strong, weak and electromagnetic forces, and gravity – attract. Dark energy, however, drives everything apart. No one knows anything more precise about it. Are we dealing with a fifth force of nature, as some physicists suspect? A more simple explanation would be that space itself causes its own growth. That would mean that space is unfolding just because it's there, like an origami flower you put in water. Albert Einstein himself flirted with this idea but then dismissed it as the 'biggest blunder of my life'. But, seen from today, the creation of space directly out of the void is much less absurd than it seemed to Einstein and his contemporaries. After all, there it no such thing as emptiness; even a perfect vacuum is full of energy, as was described in the previous chapter. So then why should this energy not cause all distances to expand and cause space to continue to grow for ever and ever? Can we not imagine the world as an origami flower folded an infinite number of times?

But why does this force, that is pushing everything apart, work in such a gentle manner? With the Standard Model we can estimate how much energy there is in a vacuum; this results in a value that is a mind-boggling factor of 10^{100} stronger than the dark energy that has actually been observed. Rarely have physicists been so far away from reality with a theory. Compared with the four known forces of nature, dark energy is effective in a dilution one could call homeopathic – like a drop of ink someone has tipped into the world's oceans. That the influence of dark energy could nonetheless be detected beyond a shadow of a doubt in the most distant galaxies, and not least in the background radiation, is one of the most exciting achievements of modern physics. There is however no chance of a yogi hovering over his rug by means of dark energy: gravity is many times stronger.

The inexplicable masses and energies in the universe may drive scientists to distraction but their existence is nothing less than a great stroke of luck for humanity. For visible matter is not sufficient to create a world that can be a home for us. Without dark matter there would be no heavenly bodies in the entire universe where life could settle. In such a world there would be no stars in the sky. There would only be gases drifting across the eternally dark expanses of the cosmos. And if the relationship of dark energy to the other forces of nature had been slightly different, even those clouds of gas would never have formed. A universe with stronger dark energy would have disintegrated so quickly that matter would have been infinitely thinly spread.

But although the effect of dark energy is weak enough to permit the formation of stable structures, it dominates the universe. Present everywhere, it adds up over the expanse of intergalactic space to immense amounts. If we take into account the fact that, according to Einstein, mass and energy are equivalent, almost 70 per cent of the present universe consists of dark energy. It is tearing everything apart, destroying every form in the course of time. And yet it is finely enough balanced to allow galaxies, stars, planets and, ultimately, us human beings to arise. This perfect balancing out of the forces is one of the great mysteries of our existence.

8

How Time Passes

*A greying beard makes you wonder why the
past can never come back. We experience
the passing of time because we are not omniscient.
The universe is growing older as well.*

Not long ago I got a shock at the bathroom sink one morning. I had returned from a long walking holiday high in the Alps, during which I hadn't bothered to shave. I was just squeezing the shaving cream onto the brush when I caught sight of myself in the mirror and thought I'd already lathered my face. In the mirror my cheeks and chin were gleaming bright, as if covered in shaving foam. Was it a hallucination? Or was my memory going already? I touched my face – it was dry, so no reason to worry about my mind. What I'd thought was shaving foam on my black hair was just a white beard.

Time passes. Nothing seems as self-evident to us as the fact that at every moment the future is turning into the past. We cannot imagine a world without time. And yet it is strange that we are so sure about it. After all, we don't feel time itself at all, we only notice that the state of something is different from the way we remember it. So could time simply be another word for change? That was what was going through my mind as I swirled the brush in the shaving cream. But in that case the philosophers who have been thinking about the nature of time for thousands of years would have been on the wrong track. My answer to the question about the nature of time was very simple: a beard turning grey

But where does the change come from? Why is yesterday different from today, why do we have to grow older? The answer is by no means obvious, for many things can be undone. Stubble can be shaved off, an impulse purchase can be returned, as a rule anyone who goes on a trip comes back home. We do, however, establish that time is passing from permanent changes. Thus we only experience the difference between past and future because there are events that cannot be reversed. A shrivelled apple will never be fresh again. A thoughtless remark can't be taken back. And the experience of life stored in one's memory makes it impossible to be 17 again.

Some things are irrevocable. That's why we experience the present as a special moment. I'll never be able to watch my daughter making her first attempts at riding a bicycle again. The sight of her wobbling and swerving around the empty car park and her cries of jubilation are just memory now. They exist only inside my head but don't seem real to me any more. The past appears to be lost forever. We may well know that there was a world at the time of the Emperor Augustus, but that world feels no more true to us than a fantasy novel. Why? Because we're not present in that world. We only see the world as real in a single moment: the 'now'.

This is by no means trivial. It's quite different with place, the 'here'. That too indicates my standpoint, but in space, not in time. But when I happen to be in Berlin I don't have thes slightest doubt about the reality of other places. Of course the world in Nairobi exists, even if I'm not there at the time. After all, I can easily make Nairobi my 'here', I just have to buy a plane ticket. But it's impossible for me to be invited to a banquet in Pompeii, not least because Vesuvius engulfed the ancient city almost two thousand years ago. That is why the world as space seems very extensive to us, while in time it's contracted to a single moment, the present.

The separation of past, present and future seems so unalterable to us that we suspect there's a law of nature behind it. But it's a law that has never been found. In fact the precise opposite proved to be the case: all basic actions in nature can be reversed. A pendulum swings from left to right then back to the left again; if a film of the movement were played backwards, no one would notice. Seen from the North Pole, the Earth rotates clockwise, but there's no law of physics that says it could not go the other way. On Venus the sun rises in the west. Mobile phones work because an antenna sends out electromagnetic oscillations but also captures them. Every atom can swallow light but also shine. And so on: no single basic equation in physics reveals anything about transience. Neither the theory of relativity, according to which the cosmos develops, nor quantum theory, which describes physical behaviour on the very smallest scale, makes any distinction between past and present. Everything that existed yesterday can be that way again tomorrow.

That was what Albert Einstein meant when, three weeks before his own death, he wrote a moving letter of condolence to the family of his boyhood friend, Michele Besso, 'Now he has departed from this strange world a little ahead of me. That signifies nothing. For those of us who believe in physics, the distinction between past, present and future is only a stubbornly persistent illusion.'

Is the passing of time really just an illusion? Of course Einstein knew why, despite all the basic laws of nature, he would never see his friend among the living again. The answer had been provided by a man Einstein admired, the Viennese physicist Ludwig Boltzmann. According to him we experience the passage of time because we know too little about the world.

At the time that was an outrageous idea. Not only was Boltzmann claiming to have come closer to solving the age-old problem of transience, his argumentation fuelled the anger of his colleagues even more. To wit, he was assuming that the world consisted of atoms. And behind what appeared to us as the passing of time were the movements of these atoms, which we could not influence. This explanation was an affront to everything the leading physicists in Berlin and Vienna considered to be proper science in 1877: a serious scientist should only talk about things that could be observed. 'Have you ever seen one?' was the general response of his opponents when Boltzmann started talking about atoms. The learned gentlemen were so incensed that one of them compared the attacks on Boltzmann to the relentlessness of a bullfight. Scientists, too, defend themselves when they are forced to abandon their prejudices.

But Boltzmann also had ardent supporters. Emperor Franz Josef received him, even wanted to ennoble him, but the scientist, who abhorred the authoritarian Austrian regime, refused the honour, saying, 'The name of Boltzmann was good enough for my ancestors, for my father, and will be for me, my children and grandchildren.' In any case, the prestige he enjoyed at court didn't help him. When the animosity of his fellow scientists did not abate, his health got visibly worse. He complained of exhaustion more and more often. Suffering from severe depression, he hanged himself in his hotel while on holiday near Trieste in September 1906. Boltzmann never lived to see his ideas gain acceptance.

That atoms exist seems to us today as much a matter of course as the fact that the Earth rotates, but even now Boltzmann's explanation of time remains challenging. It casts doubt on the unmediated nature of our perception. We perceive everything we see, feel, hear or taste as self-evident and therefore as simple – such as a beard or a loaf of bread. Things we can't touch, atoms for example, we assume are complex. But, as Boltzmann explained, the exact opposite is true: atoms are simple. If you're aware of a few parameters, you know everything about them. (The swing of a pendulum, the rotation of the planets, the radiation on the mobile-phone aerial all belong to the category of simple phenomena too.) By contrast, things get complicated with the beard and the bread. For a hair to grow and bread to taste good, a lot of atoms have to work together in a way that is anything but easy to understand. We will never be able to find out the detail of what is going on. That would not only be beyond the capabilities of our comprehension; it would be too much for any imaginable intelligence in the universe to know everything about the atoms in the loaf and the beard. It is the problem we came across in Chapter 4 with regard to forecasting the weather.

When we experience the present, this lack of knowledge is no problem. We can see the beard and taste the bread without bothering about atoms. Every object has a state that can be perceived or even measured. We automatically recognize its appearance, form, taste, pressure and temperature. The 'now' is an unmediated experience because we perceive these characteristics; we don't have to take a closer look in order to appreciate them. That is why the present is a special moment. But the states that we see, feel and taste are only the surface of the world – labels we stick on the atoms in order to be able to describe them as fresh bread, a black beard, cold and damp weather.

It's only when we want to know *how* states change that we have to look below the surface. Bread goes stale when the water inside it evaporates; beards turn white when the pigment cells on the hair follicles no longer work. That all depends on the atoms' activity, which we don't precisely understand, so that all we can do is to go by probabilities.

And by then we're already facing transience. There are few ways of preserving a state and many of destroying it. We can enjoy the taste of a slice of fresh bread now; we owe that to a certain level of moisture in the crust and crumb, to a quite specific arrangement of the atoms within it. But can we know where these atoms will be in a week's time? No, we can't. The atoms move – we don't exactly know how. Nevertheless, the bread will hardly taste good in a week's time. After all, the water molecules in the bread have far more opportunities to spread out in the air as vapour than to stay put. In all probability the loaf will dry out. The atoms are like moles in the garden: with their uncontrollable activities they ruin even the most beautiful surface.

Boltzmann's inspired achievement was to explain the difference between present and future on the basis of our knowledge and probability. We know more about the present than about the future. The bread is fresh right now, we can feel that with all our senses. Once we start thinking about the future, we lose that knowledge because we don't know how the atoms move. Boltzmann found a measurement for this ignorance that basically increases with the passing of time or at best stays the same, a measure that gives precise information about how much information on a process we lack. This measure is called entropy. Even though we cannot perceive entropy directly, the role it plays is perhaps even more important than that of energy. For entropy explains why the world is constantly changing.

Entropy increases when sugar is dissolved in tea (prior to that, we know about the sugar being in the cube, afterwards it's somewhere in the cup). It grows when apples shrivel up (some of the water from the fruit has somehow made its way into the room). It is increased when the car suddenly refuses to start ('Damn it! What's going on here?'). And it rises as we grow older.

A black beard is also a particular state. A hair can be black or blond because it is coloured by a pigment cell at the root. A huge number of chemical reactions are interacting to keep the cell alive. So if we see a black hair, we know that billions of atoms are organized in a quite specific way, otherwise the whole thing wouldn't work.

But an error can immobilize the whole system. If one single substance is lacking somewhere or other, or if just a few atoms within the genetic substance in the cell nucleus have shifted, then the cell can die off. The hair turns grey for good, without us knowing the precise reason. If something goes wrong, then everything goes wrong – that is Boltzmann's law.

When a hair turns grey we lose knowledge of the system and entropy grows: the beard has passed from a more improbable state to a more probable one. After all, the pigment cell has only very few ways of performing its function, but very many of failing. To get six numbers right in the lottery is highly unlikely, for there is just the one combination of numbers for that. The chances of matching four are better because more combinations are possible. In biological terms a black beard corresponds to matching all six numbers in the lottery.

Can a beard that's turned grey become black again? Theoretically it could. The laws of nature always allow events to reverse. However, conditions in the cell are extremely complicated. How probable is it that thousands and thousands, perhaps even millions of atoms will spontaneously combine in such a way that the cell can produce black pigment again? That would indeed be an incredible fluke. The patience one would have to show in waiting for the colour to return cannot be measured in human lives – the life-span of a universe isn't enough for that.

The improbable states vanish to be replaced by more probable ones. The more improbable a state, the more transient it is. The more probable a state, the more stable it must be and therefore the less likely it is to revert. Where there is a forwards but no backwards any more is where we experience the passing of time.

If we could turn back every process in the universe, time would not be of any great importance for us. We would then be able to restore everything to the way it once was and be unable to distinguish past from present. An all-powerful being would indeed exist outside time. We, however, are imprisoned in it because we can't influence, or even predict, most events.

Behind the puzzle of time lies chance – the unpredictable dance of the atoms. This connection explains why Einstein felt bound to see the division between past, present and future as an illusion. He believed there was no such thing as chance. Einstein was a determinist. He was convinced that everything that happens in the world is determined by the laws of nature. It is only because we lack complete understanding of the course of these events that it seems to us as if time were passing.

It is difficult to argue with a determinist, whose perspective is that of an all-powerful being. But what would be the advantage of following this view? After all, we want to understand the world as humans, and being human happens to come with a large degree of ignorance. Thanks to Boltzmann, we are now at least able to see this ignorance more clearly: the greater the entropy, the more information we lack, the more difficult is is to revive the past and the more we are subject to time.

If we are unable to exert any control, it is to be expected that things will go from the improbable to the more probable state, but not the other way round. The particular will be displaced by the more general, which is why yesterday has gone for good.

We experience time as decay wherever we leave lifeless things to their own devices: bread loses its freshness, buildings deteriorate, erosion wears down mountains. Living beings resist their decline, but not for long. In any case they destroy order in the course of their existence. The fact that entropy can never decrease is nothing other than a mathematical formulation of that experience. Physicists call this the 'second law of thermodynamics'. If you take all objects involved in a process together, their entropy can never drop. That is the second law. In the long run what is more improbable makes way for what is more probable. Not one single process that goes against this law has ever been found.

Everything in the universe has to grow older, even the universe itself. We must assume that in the distant future the sun will go out, the planetary system will disintegrate, even the galaxies will disappear. After all, the present cosmos, in which stars shine and planets revolve, is a quite particular arrangement of matter. And what is particular is transient. Much more probable is the greatest conceivable confusion.

But then again everything surrounding us is particular. The sugar in our tea, the apples, the people, the stars – how can their existence be reconciled with the laws of probability? What is particular is never lasting. In the long term the greater probability, chaos, always wins. The secret of time leads us to what is clearly an even greater puzzle: how is it possible for there to be structures in the world? Why could sugar, apples, stars arise? Why are we here at all?

A black beard is already so improbable that you wonder how it can exist at all. The interplay of the atoms in the pigment cells at the root of a hair is extremely delicate and, as any disruption can paralyse the whole system, one would be more inclined to expect the cell to fail rather than dutifully continue to produce its pigment. The pigment cells do, indeed, constantly die off, like almost all cells in the human organism. But the body can replace dead cells with new ones. We rejuvenate ourselves. That is the sole reason why we don't go grey when we're very young.

So do we then manage to outwit entropy, at least in the short term? By no means. We only suspend our own deterioration by creating even greater disorder elsewhere. We eat and breathe. Bread and apples are more ordered states than human excretions. In order to renew a few hundredths of a gram of cells per day, we send several kilos of food and water down the toilet. Moreover our organism uses up a good 1,500 litres of oxygen a day in order to transform food into carbon dioxide, water and warmth. In so doing we turn a hundred thousand times more matter from a particular into a more general arrangement, from a more improbable into a more probable state, than we combine into ordered structures. With every day we spend in the world, we increase the entropy in our environment.

Planting a little apple tree doesn't make things better either, because the result is even more unbalanced if we include the production of food and oxygen. We owe both of these to plants that have converted carbon dioxide and water. But the growth of leaves and fruits also has its price, only we don't notice the disorder caused by the plants the way we do processes in our own body. Entropy also arises during photosynthesis, which turns sunlight into biomass and heat. This might seem surprising at first: why is heat less ordered than light? Both forms of radiation consist of photons, but they are differently constituted: at the same intensity, light contains few high-energy photons, while heat contains many low-energy photons. That means that heat is a less ordered state. Sunlight and heat radiation are like singing and bellowing – there are few possibilities of turning sound into a melodious song, but countless ways of creating cacophony.

Sunlight comes to Earth from outer space, heat goes back there. If the Earth did not give off heat it would have become a desert planet long ago. Our own life makes entropy on Earth grow, the life of plants produces it in the solar system.

But if heat is more improbable than light, why then is there light? Photons are formed in the sun during the fusion of hydrogen nuclei and are sent out into space, never to return – yet another irreversible process. The Sun, too, is growing older. Its mere existence is really a miracle, much more astonishing than a black beard. The Sun only exists because the whole of space is in a very improbable state. The cosmos contains huge masses, masses attract each other: why then didn't matter implode long ago? Why is there a shining star in the middle of our planetary system rather than a black hole?

There is only one possible answer: because long before the emergence of the Sun, matter was in a fantastically improbable state. When matter developed, it was spread so evenly over the young cosmos, so well-ordered, that it defied gravity. This is revealed to us by the cosmic background radiation that was emitted when the atoms were formed, 3,000,000 years after the Big Bang. Because space is expanding like a balloon and moving all masses further and further away from each other, matter didn't agglomerate into black holes. It is solely because of this that clouds of gas and galaxies, stars and the Sun, plants and black beards could develop.

An abyss of improbability opens up. The amazing fact that we exist can only be explained by even more amazing conditions in earlier times. And the further we look back into the past, the more improbable the states become.

Attempting to understand why a beard turns grey takes us to the beginning of the world. And this beginning seems quite different from the way we usually imagine it: the present universe did not arise from a probable state, that is to say out of chaos, but out of a truly unimaginably improbable one. Chaos would have ended up in black holes. In the beginning, there was order everywhere.

At first sight we are disturbed by this cosmology, because it is not only contrary to the creation myths of most cultures but also to our intuition. One objection is obvious: everything in the universe that seems ordered to us – galaxies, stars and planets, life itself – only developed over the course of history, in a process that lasted billions of years. But this objection is misleading. The order we perceive today is just the pale reflection of the much greater regularity that must have reigned at the very beginning. Without this initial order, the structures that make the world seem organized in some places today would never have arisen. They only formed at the cost of an increase in disorder elsewhere. Overall entropy increases with time, the cosmos as a whole develops from an initially extremely improbable to a more probable, and therefore less ordered, state.

The Oxford mathematician Roger Penrose once worked out how improbable it was that a cosmos like ours should arise. He came to the incredible odds of one chance in $10^{10^{23}}$. This figure is so small that a piece of paper would have to be many light years long and wide to print it out at a normal size. This huge sheet would be entirely covered in zeros after the decimal point. Only someone who went to the trouble of travelling to the end of the last line in a spaceship would eventually find a different digit there. The degree of improbability of the birth of our universe corresponds to the degree of order at the beginning.

But where did this order come from? At present no one dares speculate about that, even the boldest theoreticians remain silent. The origin of the order to which we owe everything is completely inexplicable. But my beard points to this very first of all secrets. And that has reconciled me to the fact that it is turning grey.

9
Beyond the Horizon

The night is dark
because the world had a beginning.
Since then the universe has been expanding.
Space is bigger than we can imagine.
Thoughts on being amazed.

Why is the night dark? When one of my children asked me this question as I was putting her to bed, I replied, as if it were obvious, 'Because the sun isn't shining.' But my daughter was persistent. She said that the stars give off light as well.

'The stars are a long way away,' I said. 'That's why their light is weak.'

'But there's so many of them,' she said.

'But space is very big.'

'And full of stars, yes?'

'Yes.'

'Then the sky should be bright at night.'

'Go to sleep now,' I said, amused at the thought that a seven-year-old just has no idea of the dimensions of the universe.

But in fact, I was the one who was wrong. The question my daughter asked has been bothering astronomers for centuries. Why it goes dark at night was already a mystery to Johannes Kepler, who was the first to calculate the orbits of the planets in the early 17th century. Kepler had realized that in infinite space, which had since time immemorial been furnished with evenly spaced-out stars, there ought to be no darkness. In whichever direction you look, there's always a star shining somewhere. You can't see through a forest for the same reason: there are gaps between the trees, of course, but the larger the forest, the more trees there are blocking your view. In just the same way, more and more stars appear in the sky the further we look into space. In a cosmic space of infinite depth, the whole sky would be covered in stars. Such a sky would shine as bright as the sun.

So the night sky tells us that the cosmos we can see must be finite. It's either of a finite size or a finite age, or both. We find one as difficult to visualize as the other. If the universe stops somewhere what, then, is there beyond its boundary? And if it began sometime or other, what happened before its beginning? A finite world seems to us to be intrinsically contradictory. We would, of course, find the opposite equally baffling. Infinity is also beyond our comprehension. Either way, we have trouble understanding reality.

When we become aware of this inability, we are amazed. We feel overwhelmed by a reality that is quite different from what we imagined – or could imagine. Although we cannot understand this reality, we feel bound to it in some inexplicable way. And yet this amazement is more than a mystical feeling; it derives from our rational thought, after all. We are amazed when an unexpected connection dawns on us. We find the answer to a question and, much to our amazement, ten further questions immediately arise. Amazement is the transition from ignorance to deeper understanding: the world opens up to us. And that is one of the main reasons why we most often see it on a child's face.

For Albert Einstein these moments of amazement were the most precious of all. 'The most beautiful experience we can have is the mysterious. It is the fundamental emotion that stands at the cradle of true art and true science. Anyone who is unacquainted with it and can no longer marvel, no longer be amazed, is as good as dead, his sight gone.' So said the great physicist in 1932. He read these lines for a recording he called 'My Credo'.

Being confronted with something unfathomable is an experience we have less and less often. In the eyes of most of our contemporaries, the world has lost its secrets. And if there is a question they can't answer, they trust that there is some expert who can. That is understandable. Every day we hear announcements of new breakthroughs in scientific research. Our dealings with the marvels of technology in particular encourage us in the assumption that the human race has understood how nature works. Space travel, high-tech medicine, faster and faster computers in everyone's pocket – everything seems possible as long as the will is there. (Only now and then does a cold remind us that our dominion over nature still has its limits.)

However, we do tend to avoid the ostensibly simple questions that children ask and to which Albert Einstein owed his deep insights: why is the night dark? How did the world begin? What is time? We answer our children in a few dry words: because the sun's shining on the other side of the world. With a big bang. What you can see on the clock. This superficial response robs us of the stimulating experiences Einstein describes: amazement at the world in which we live. And the joy of discovering connections within it.

Our amazement seduces us into looking for hidden connections and thereby to even greater amazement. When, for example, we look up at the night sky with the naked eye, it shows us the traces of the Big Bang. And when we reflect on that we're immediately faced with questions such as whether the universe can be finite or has to be infinite. 'The eternal silence of these infinite spaces frightens me,' the 17th-century French philosopher and mathematician Pascal wrote, doubtless after such an experience. He could have had no inkling of the Big Bang. Was he right all the same? Is the universe really infinite?

Today we know that the night sky is dark because the cosmos wasn't always there. We see a finite number of stars, not an uninterruptedly illuminated sky – which we would only be able to see if the light from an infinitely great number of stars were to reach us. Whether there is an infinite number of stars is something we don't know, but even if there were, we wouldn't be able to see all of them anyway because the universe has not existed forever: the emissions from the most distant heavenly bodies simply haven't had enough time to reach Earth. According to the latest data, the Big Bang happened 13.8 billion years ago, therefore only stars whose light could reach us in 13.8 billion years can appear in the night sky. But since there is only a finite number of those, the firmament stays black. Pascal could have had no idea of this connection between the speed of light, the darkness of night and the beginning of the world.

Anyone using a powerful telescope will discover a further indication of the beginning of the universe: distant stars shine red. Their colour indicates that they are moving away from us because the universe is expanding. The American astronomer Edwin Hubble discovered this phenomenon in 1929. It's explained by the fact that light is a wave. When the space in which light is spreading expands, everything inside it is stretched – and that includes the wavelength of light. And as a result, the colour changes, because each wavelength corresponds to a colour. We see short waves as blue, long waves as red.

The further away the stars we look at are, the more their light has shifted towards the red end of the spectrum. That is to say, the more the space between them and us has already expanded. This effect appears irrespective of the direction we look in. All the distant stars are zooming away from us, therefore the whole of the cosmos must be expanding. The most distant of the galaxies known today are moving away from Earth at the unimaginable speed of 7,000 million km/h (4,350 million mph).

If we could rewind the history of the universe like a film, we would see the opposite movement. The heavenly bodies would come closer and closer together and we would reach that point in the distant past when all the things we can see today were concentrated in one patch. That point in time is the Big Bang. The outward movement of all the stars proves that there was in fact a beginning.

Many people imagine the universe at the time of the Big Bang as a ball of fire. The story of creation then goes roughly like this: something exploded with unimaginable violence and ever since, driven by the force of this explosion, the universe has been expanding. Consequently, we're sitting inside an inflating ball that is assuming increasingly immense proportions.

But can it really be like that? If the idea of the universe as an expanding ball were correct, then the cosmos would have a boundary – the cover of the ball. And that is impossible. Because in that case, there would have to be an outside; the universe wouldn't be everything.

A universe with external borders is unthinkable. Does that mean we live in an infinite cosmos? Not necessarily. There's a second possible way of avoiding the idea of a universe you could fall out of: perhaps the universe is curved in on itself.

Space is elastic, it can bend and twist like rubber: that was the brainwave that gave Einstein the idea for his theory of general relativity. Heavy masses distort their surroundings – like a stone making a dent in a taut rubber membrane. In this way, masses divert light, as explained in Chapter 7, giving rise to spooky images in the sky. But space can be warped not only in the vicinity of a star or a galaxy, but as a whole, just as you can distort a lump of rubber if you tug at it for long enough.

What does space warped in such a manner look like? Let us imagine an ant walking around on the outside of a ball. Its world is a surface, namely the outer surface of the ball. The ant is therefore only moving through two dimensions, not three. But the surface it is crawling over is curved. Although the ant thinks it is going in a straight line, in reality it is constantly changing direction. It doesn't notice that, however. The ant could keep wandering around the ball forever without ever hitting an edge. The surface is boundless and yet finite. If the insect keeps going it will eventually come back to where it started out from. If the ant is a particularly clever beast, it might perhaps be surprised by that. It will however never become aware of the actual nature of the surface on which it is walking. Its mind is only equipped for a life in two dimensions.

It would be quite possible for us to be living in a warped three-dimensional space without realizing it. In our everyday life, the warping just wouldn't be noticeable. It would be beyond our powers of imagination, just as the insect is incapable of visualizing the true shape of its home.

That, even having reached this point, we don't have to give up on exploring the world, is one of the greatest triumphs of humanity. Admittedly we only see a tiny part of reality, as Albert Einstein once wrote: 'Nature shows us only the tail of the lion, but I have no doubt that the lion goes with it, even if it can't reveal itself to us directly because of its immense size. We only see it in the way a louse sitting on its back sees it.'

But unlike an insect, we know how limited our perspective is. We also have a special language we can use to describe circumstances that are beyond our vision: mathematics. There is no problem expressing the warping of a three-dimensional space in that language. And measurements can be used to check what the mathematical formulas predict.

It was only at the beginning of the new millennium that we managed to measure the warping of the universe. It was a challenge, because the larger the object, the more difficult it is to see its shape. Even the spherical form of the Earth is normally beyond our perception. With a radius of more than 6,000km (almost 4,000 miles), the curvature is minimal and, moreover, the horizon distorted by irregularities. You have to be able to see a very long way to establish that the surface of the Earth is curved. Only someone who is watching the masts of a distant ship rise on the horizon of the ocean, or is zooming across the stratosphere in a spaceship, can tell that the Earth is not flat.

The dimensions of space demand much more extensive vision. Without technology the shape of the universe would remain unknown, but in 2001 the Americans launched the Wilkinson Microwave Anisotropy Probe (WMAP). Eight years later the Europeans sent an improved version, the Planck space probe, up into the cosmos. These two probes measured the background radiation which, as the first flash of light after the Big Bang, has been heading towards us for 13.8 billion years. There can be no older radiation than that, therefore it denotes the extreme horizon, the edge of the universe that is visible to us. The two space telescopes looked out as far as it is possible to see from our solar system.

The two probes were built in order to see what you might call the ship appearing over the horizon, that is, to establish the curvature of the cosmos. The first mission, however, did not produce a result: the data recorded by the American telescope did not show the least sign of curvature. Was the probe too imprecise in the way it worked? All hopes were pinned on the next space telescope. The European measuring instruments, with far greater power of resolution, would perhaps deliver different results. But Planck also reported 'curvature zero'. The universe appears as flat as a matchbox.

A flat universe can only be infinite. Otherwise it would have edges at which things could fall out. So do we live in an infinite matchbox? That would indeed be the simplest explanation for the results of the two space probes. We can't be certain, however. For one of the most remarkable facts established by modern cosmology is that although we can date the beginning of the universe down to a few million years, we are not able to establish its size. The true measurements of the universe remain fundamentally hidden from us. The data from the Planck space probe, published in 2013, only tell us the minimum size of the universe, because of course we only receive light from the nearer regions of the universe – the radiation that has had time to reach us since the Big Bang. We cannot see those parts of the cosmos that are further away.

It is conceivable that the finding of a flat universe results from an error in measurement. Perhaps the cosmos really is curved, but only to such a minimal extent that the Planck probe is unable to detect such a tiny distortion. In that case, we could hold on to the idea of a universe that is finite, but boundless because of its curvature. However, since Planck operates with the greatest precision, the curvature it overlooked can only be extremely small. And the situation with the cosmos is the same as with our terrestrial sphere: the smaller the curvature is, the greater the diameter must be. That means that a universe whose curvature we cannot measure with our present technology must be of huge dimensions. It would be as good as infinite.

A further abstract solution to the problem of a finite but boundless universe would be by means of complicated geometry. In a cosmos like that, it would be as if all the exits from the matchbox were connected with the entrances in such a way that it is impossible ever to leave the box. Higher mathematics does actually allow for structures that are at the same time both flat and infinite. The simplest shape that comes into question is known as a hypertorus and corresponds to the three-dimensional surface of a bicycle inner tube fitted on a four-dimensional wheel; surely even the imagination of an Albert Einstein would give up at that. Other possible spaces are interwoven, convoluted, twisted in even more fiddly ways. If we did live in such a universe, it would make a mockery of our longing for a reality that is readily comprehensible, perhaps even elegant.

Cautious estimates come to the conclusion that hidden beyond the horizon of the universe that is visible to us is a cosmos that is at least 250 times larger. Most cosmologists, however, assume different dimensions. A somewhat less conservative analysis of the data from the American WMAP space telescope results in a volume that is 60 billion times larger. Joseph Silk, an astrophysicist at Oxford University and a pioneer in the investigation of the cosmic background radiation, even assumes that the cosmos must be one googol, 10^{100}, times larger than the part of the universe whose light reaches us.

This means that if the universe were an ocean, we wouldn't set eyes on even a single drop of it. That is how narrow the limits set on our understanding are in a limitless universe. The star-studded night sky seems immeasurable, but it is vanishingly small measured against what is behind it the cosmos that is inaccessible to us, from which no information reaches us and never will. Pascal was right to talk about the 'eternal silence' of immense spaces.

The universe that is visible to us has a radius of 46 billion light years. That corresponds to the distance that light could travel in the 13.8 billion years since the Big Bang. This distance is greater than 13.8 billion light years because the universe has expanded in the meantime. We can neither see nor reach anything that is further away, because light from regions beyond the horizon is still on its way to us. And because the cosmos is still expanding, they will remain invisible to our descendants as well.

Let us assume for the moment that the same laws of nature operate everywhere in the cosmos. Why shouldn't they? None of the astrophysicists' measurements provide any indication at all that there is some fundamental change with increasing distance from Earth. Our planet is not a special place in the cosmos, so we can at least make a well-founded assumption about the universe beyond the horizon: things there are basically the same as they are here. There are galaxies, stars, planets.

The bigger the universe is, the more heavenly bodies there will be hanging around in it. And the more planets there will be among them that are similar to Earth. You just have to imagine a simple game of dice to see why: the longer you play, the more likely it is that particular numbers will come up at some point. That there has been no six after ten throws can happen, but only rarely. After all, the probability of a six coming up at some point in a series of ten throws is 84 per cent. The chances are even better after twenty: 98 per cent. If the game continues, the chance of a six being thrown approaches 100 per cent: after 50 throws the probability is 99.99 per cent, after 100 throws even 99.9999999 per cent. After 100 throws, the six is almost certain to have come up once.

It's exactly the same in the universe: the bigger it is, the more probable it is that certain things will happen somewhere. How credible is it, for example, that there is a planet revolving around a distant star on which life such as we have on Earth would be possible? That a twin planet like that should turn up in our cosmic neighbourhood is conceivable, but highly improbable. The chances of it happening in our galaxy, the Milky Way, are already better. In February 2017, NASA announced that a system with seven planets similar to Earth had been found in the constellation Aquarius, about 40 light years from Earth. It is true that the radiance of the central star, a red dwarf, is much weaker than that of our Sun, but life does at least seem possible on some planets in that system. The more areas in the cosmos this search is extended to, the better our chances are of actually finding something like a second Earth. The probability grows at an exponential rate as the search area expands, just as after a long session of throwing the dice, a six is bound to come up.

Let us carry on with the game. Anyone who doesn't want to get just one six, but two sixes in succession, needs more patience. After 170 throws that has a probability of 98 per cent; for a 99.9999 per cent chance it takes 500 throws; 99.999999 per cent is reached after 700 throws. By the same token, we could demand more of our sister planet than merely similar living conditions to those on Earth. Let us then require that two-thirds of the surface of this other world is covered in oceans, or perhaps that exactly 14 peaks over 8,000m (26,247 feet) rise up from it. Even if the Earth should have no such double in our entire visible universe – the cosmos is much, much bigger than that. And with its expansion, our chance of a hit increases.

As long ago as the 16th century, Giordano Bruno, a Dominican friar and natural philosopher from Italy, had realized that in an infinitely large universe, intelligent life must have arisen an infinite number of times. He paid with his life for his reflections on the cosmos. The Inquisition pronounced him guilty of heresy and had him burnt at the stake. It is said that when Bruno was led to his execution on the Campo de Fiori in Rome, the guards had screwed his tongue into a wooden clamp, in order to prevent the visionary from speaking to onlookers from the pyre. There is no doubt that the Vatican saw the monk-turned-natural-philosopher as an extremely dangerous man. The Church pursued Bruno for almost twenty years before he was finally captured in Venice. And their quarry was well aware of the effect of his teachings. It is recorded that after the guilty verdict, he replied to the judges, 'Perhaps you pronounce this sentence against me with greater fear than I receive it.' Until 1965, Bruno's works were on the Index of books forbidden by the Catholic Church.

The Vatican had good reason to fear Bruno, especially as he argued along theological lines: the infinite, incomprehensible God, he said, could only have created an infinite, incomprehensible universe. Anything less would not be worthy of him. But an infinite number of suns, an infinite number of planets and an infinite number of lives do not fit it with the teachings of the Church, that humanity's history of salvation was unique.

However despicable it is to muzzle an awkward thinker, however repulsive the methods of the

Inquisition were, we can understand the Church's concern. Bruno's theses could have set off immense unrest. Even today, the idea that the universe might be infinite is disturbing. An infinitely large, uniform universe would not only nullify the uniqueness of humanity and its history, it would not admit of any uniqueness at all.

In such a cosmos it is not only inevitable that life exists on an infinite number of worlds, there could also be several planets out there on which there is a rainy country in the northern latitudes whose inhabitants are proud of their tea and their royal family. An infinite universe can even contain an infinite number of such planets. And on every one of these planets, there would not just be a Premier League, pizza and Coca-Cola, but also a duplicate David Beckham and copies of all the people living on Earth. With a probability close to certainty, each one of us has an infinite number of doubles in the cosmos.

In the infinite universe, absolutely everything the laws of nature allow will take place. On one planet, all the murders solved by Sherlock Holmes have happened before, on another, there are hobbits. There will be heavenly bodies on which silicon-based intelligent life is flourishing, perhaps even consciousness that is generated by whole galaxies.

We cannot know whether we actually live in such

a reality. But there is much to suggest that we do. There is no doubt that we live in a universe of vast dimensions, of which we see only a tiny part, and beyond which are huge spaces that are inaccessible to us. That these spaces might be infinite is not fantastic speculation. On the contrary, a cosmos that extends endlessly in all directions would be the simplest solution to the puzzles that a look at the black night sky presents us with. We will see why in the next chapter.

Naturally we are bewildered by the idea that every person could exist in an infinite number of versions. That man over there, on a planet billions of light years away, who's sipping his green tea and finishing the penultimate chapter of a book on the universe – am I that man? But even that shock would pass. Again and again, science has confronted humanity with insights that we initially rebelled against but eventually accepted. It is therefore quite possible that my descendants will not be able to imagine a life without their doubles, that they will wonder how people could ever have put up with the loneliness of being unique individuals.

10

Why we exist

*In each of us one of the most astonishing
characteristics of the universe appears:
intelligent life is not only possible but even probable.
How can anyone maintain therefore
that we are meaningless?*

A concatenation of exceptional occurrences led to my birth. And this series of events began long before my parents' eyes met, for it is more than astonishing that there was a woman at all who could bring me into the world. I have a gigantic meteorite to thank for that. When it crashed into the Earth, dust and ash darkened the Sun for months on end; three quarters of all species disappeared for good. If, 65 million years ago, the meteorite had entered the Sun's gravitational field at a minimally different angle, it would have missed the Earth: in that case would it have remained a planet of the dinosaurs? But by plunging into the Gulf of Mexico, the meteorite made room for unremarkable, nocturnally active creatures who bore their offspring in the womb – our ancestors.

The Moon also made its contribution to my appearance on the scene. With its gravitation it secures the rotational axis of our planet in space, otherwise the Earth would wobble. Without the

Moon there would be no regular seasons, the climate would be chaotic. Plants and animals would presumably never have developed on Earth. Would life as we know it have had any chance at all on it? Probably not. There is reason to believe that the first single-celled organisms developed in the ocean's tide pools, in which they were exposed alternately to the heat of the Sun and the water. Without the Moon there would be no tides, without their ebb and flow no human beings.

And where did the Moon come from? The fact that our small planet has such a companion is phenomenal. The Earth would never have been able to capture the Moon by itself with its own gravity. Instead it was a huge collision that gave it its satellite. Shortly after Earth was formed 4.5 billion years ago, it collided with a protoplanet called Theia. It is presumed that Theia was the size of Mars. The Moon came from the debris of the crash. Needless to say, such a coming together is extremely improbable in the vastness of the planetary system. It seems even more improbable that the two hit each other in such a way that the collision sent a lump the size of the Moon flying off into space. But clearly that is precisely what happened.

It would be possible to go on for page after page with the story of the strange events that led to my birth.

The arrangement of the neighbouring planets in the solar system, the development of our planet's strong magnetic field and continental drift would be among them. And the way life arose from dead matter: large molecules had to form cells and combine in such a way that they could propagate themselves. The chance of that happening unaided would be comparable to the possibility of a tornado sweeping across a junkyard and making a jumbo jet out of the parts lying around, the British astrophysicist Fred Hoyle once scoffed.

Each single one of these events seems surprising in itself, but I owe my existence to not one but to a long sequence of special occurrences, from the birth of the Moon out of a collision to the extinction of the dinosaurs at just the right time. Otherwise, human beings would never have populated the Earth. When we review the course of events on Earth, from the genesis of our planet out of a cloud of dust to our own appearance, we can hardly imagine all that happening spontaneously. We automatically wonder about the hand of a director organizing it.

We find it difficult to imagine that life should have begun spontaneously on our planet alone. The very idea gives us the shivers, because in that case our existence would be a mere curiosity, arisen by chance: a bit of hustle and bustle on a pale blue spot somewhere in the cosmos – interesting but absurd, and presumably soon forgotten. Was it just a whim of the universe that four billion years ago on the edge of the Milky Way, dead matter came to life – once, but never again?

That is precisely the view of life that leading scientists have taken until recently. The French molecular biologist and Nobel prizewinner for medicine, Jacques Monod, representing the views of many of his colleagues, described it particularly elegantly. He pointed out the well-nigh incredible chance nature of the development of life on Earth. Then he concluded, somewhat grandiloquently, that man was alone 'in the unfeeling immensity of the universe from which he emerged only by chance'. The universe, it turns out, was not pregnant with life.

Monod's book, *Chance and Necessity*, was published in 1970 and immediately became a bestseller. In almost every line you can sense his bewilderment at the complexity of even the simplest organisms. Monod is amazed at the secrets of the world inside the cells that he was among the first scientists to penetrate. This wonderment makes Monod's conviction that life is a unique accident all too understandable.

But while he correctly recognized the complexity of living organisms, Monod far underestimated the complexity of the universe. At that time astronomers lacked any idea of the true size of the cosmos. Most suspected that we lived in a finite space a few million light years in diameter. They only knew of one solar system, ours. Giordano Bruno's theory of an infinite universe filled with an infinite number of planets was for them just the wild speculation of a heretic from a pre-scientific age. In short, people thought the universe was much smaller than it actually is.

In a small cosmos, life would indeed be difficult to explain. If there was only one planet that could develop life, it would have as good as no chance at all. Therefore Monod was bound to see our existence as the result of an accident.

Today we know we are surrounded by a universe of immense proportions. And ever since the Kepler Space telescope, launched in 2009, began scouring the sky, dozens of new planetary systems have been discovered in the Milky Way on a daily basis. Quite a few contain planets that are similar to Earth. A cautious estimate suggests that in our galaxy alone, there are at least 100 billion planets orbiting the stars. As the visible universe contains at least 100 billion galaxies, that brings us to the huge number of 10^{22}, that is ten thousand billion billion, planets. And that is just the number of heavenly bodies in the tiny part of the universe we can observe. The parts of the cosmos that are hidden from us must contain far more planets.

These numbers change everything. For however small the chances of a particular planet bringing forth life may be, the more planets there are, the more likely it is that something like that will indeed happen on another planet. Even the most improbable events will come to pass, given enough chances. That is what the law of large numbers demands.

Monod considered the prospect of intelligent creatures on just a single planet, Earth; from that he concluded that such life in the cosmos was utterly improbable. But more important than the probability of a single event happening is the frequency of the attempts made. If I won the jackpot in the lottery, that would be amazing. But if a million people tried their luck, it would be amazing if no one had picked the right six numbers. And if billions of lottery tickets were entered into the draw, people would be surprised if there weren't thousands with the right six numbers. Monod did not take into account that repetition will offset any probability, however small. With ten thousand billion billion planets in the visible universe alone, it would be incredibly bad luck if intelligent life had not developed on at least one of those planets.

The American astrophysicists Adam Frank and Woodruff Sullivan have worked out what 'incredibly bad luck' means. Roughly every fifth sun-like star in the cosmos has one or several planets in the so-called life-friendly zone, where there is the possibility of liquid water. Frank and Sullivan's formula combines the chances of intelligent life arising on a planet in the life-friendly zone with the probability that the cosmos as a whole has produced intelligent creatures. The result: provided that intelligent life develops on planets in the life-friendly zone with a probability higher than 10^{-24}, a millionth of a billionth of a billionth, then it is to be expected that it exists in the visible universe.

This threshold is exceptionally low. We just have to remember how quickly life on Earth got under way. A good 500 million years after its birth from a cloud of dust, volcanic activity and meteorite impacts had abated to such an extent that there was a chance of life on Earth – and it emerged. The oldest known single-celled fossils date from this very time. This quick start suggests that the probability of the development of life under suitable conditions was by no means too small – at least distinctly greater than 10^{-24}.

We don't know what the chances are of intelligent beings developing out of microbes. The probability is presumably very small, but not negligible, as shown by recent research results in behavioural science and neurobiology. The first signs of intelligence can already be observed in supposedly primitive creatures. Humboldt squid communicate by changing the colour of their skin, allowing them to coordinate their hunting activities; octopuses can learn to open screw caps with their tentacles. Bees with their tiny brains have a memory and use a type of symbolic language when they tell each other with their waggle dance where nectar can be found. Crows, direct descendants of the dinosaurs, can use tools, recognize themselves in the mirror and learn to count. It therefore seems plausible that Earth would be inhabited by creatures with understanding, even if there had never been humans or even mammals. In that case, perhaps cleverer descendants of the birds would rule the land and highly developed cephalopods the oceans.

So the probability of intelligent life being able to establish itself on our planet is probably well above Frank and Sullivan's threshold value of 10^{-24}. Consequently we must proceed on the assumption that the universe would more or less inevitably have generated intelligent life. And since the cosmos, as far as we are able to observe it, is uniform, there is no reason why that should only happen on Earth. We must assume that intelligence exists elsewhere in cosmic space: creatures that can recognize themselves and their environment, that make plans, pursue goals. We are not alone.

It's an exciting finding. But it is the way we came to this conclusion that must give us pause for thought:

we can deduce that we are not the result of a cosmic accident solely from our own history and the nature of the universe. Our existence therefore reveals one of the most astonishing qualities of the cosmos: intelligent life is not just possible, but even probable. Can anyone then still claim that we are meaningless?

There is no question that the probability that we would appear on planet Earth of all places, on the edge of the Milky Way, in the vicinity of the star Alpha Centauri, was very small. But it had to happen on one of the vast number of planets. The universe was definitely pregnant with us.

Why is the universe arranged so that it produces beings like us? Religious people explain the order of the world as the work of a benevolent creator. We can't argue with them. It is impossible to prove either God's presence or His absence. But a scientist, even a religious one, is not satisfied with such an answer, and their research is the attempt to understand the world in terms of its natural causes. A scientist who invokes higher powers has given up.

Albert Einstein, never short of an ironic turn of phrase, asked whether God had a choice when he created the world. He devoted more than half his adult life – the last three decades before his death – to that question. 'What really interests me,' he told his assistant around 1945, 'is whether God could have made the world different.' He suspected that the answer had to be 'no', because Einstein was convinced that nothing in nature can be arbitrary. To his eye, the world seemed logical, understandable, simple and beautiful. Each of its laws had its own justification and worked together with the other laws. And a God who was worthy of the name would not disrupt that order with wanton interference.

What form does this order take? And can we understand its deeper connections? Einstein directed all his efforts to the search for an ultimate theory that would answer such questions. It was a course on which he was completely alone. His colleagues preferred to look at problems that seemed more capable of solution.

What Einstein was dreaming of is now often called the 'theory of everything'. The name is misleading, for such a formula would not allow one to work out all events in the world – that is, as we saw in Chapter 4, fundamentally impossible. Nature, ramified into ever finer branches, is beyond complete computation; we can't even provide reliable weather forecasts, despite knowing the equations the weather obeys. But a theory in physics is not a photographic record of reality, but rather something like a blueprint for the world – and no one would expect the plan for a house to show every nail on the staircase.

The 'theory of everything' Einstein longed to find would be like the blueprint for the universe. It would explain how reality is arranged. From it we would be able to derive the framework of the laws of nature, the structure of the world that allows our existence. The theory would, for example, have to explain the fact that space has three dimensions, not two or four. It would also provide information about the size and shape of the universe. We would finally be able to understand ourselves and our place in the cosmos.

But Einstein's lonely search brought no success. He kept making fresh attempts to find an explanation for nature that would be more comprehensive than the physics that was known. And he kept on rejecting his own ideas. 'Most of my intellectual children end up very young in the graveyard of disappointed hopes,' he wrote in a letter of 1938. When he died in 1955, he had not come any closer to this 'theory of everything'.

One reason why he was bound to fail was that his theory tried to unite only two of the four elementary natural forces known today, electromagnetism and gravity. Einstein left out the strong and the weak forces that ensure the bonds in atomic nuclei, which were largely unresearched at the time. But without those forces, it is impossible to understand how matter can assume its form and mass, why there are various elementary particles and why the stars shine. Einstein was like a cartographer who wants to draw a chart of the Atlantic Ocean but just knows the coastlines of Europe.

Is the universe necessarily the way it is? It was impossible even to make a start on solving such a puzzle with the knowledge of the 20th century. If God made a choice when He created the world, there was no hope of finding out what He had been up to. But even though no one took his efforts seriously, Einstein had, as so often, asked the right question.

For decades, Einstein's lonely search for an order that explained everything was as good as forgotten – until cosmologists began to wonder why the universe is on the one hand so big, and on the other so uniform.

Wherever you look in the night sky, you see the same thing: the far-distant galaxies and stars are so thickly and evenly spread, it's as if someone had carefully sprinkled them over the cosmos with a salt cellar; nowhere are there darker or lighter patches. Even more impressive is the regularity of the cosmic background radiation. This remnant of the first light after the Big Bang comes to us from all directions with almost the same wavelength. The variations shown by the spot pattern of the celestial map for this radiation are less than one-tenth per thousand. Clearly the universe, as far as we can observe it, not only looks the same everywhere today, it has always been homogenous.

The uniformity of the cosmos suggests that Earth is not a cosmic exception. Everywhere there are planets and stars in their orbits; everywhere is governed by the same laws of nature; everywhere the same conditions also applied in the past. And that is certainly more than astonishing. The same conditions can easily be established in a neighbourhood, because neighbours know about each other and exchange information. But how should the eastern night sky know about the structure of the western night sky?

The distances in the cosmos exclude any such exchange. They make the transfer of information not just practically but fundamentally impossible. If the light from a galaxy ten billion light years away to the east and that from another galaxy ten billion light years away to the west reaches us, then these two galaxies are 20 billion light years away from each other. But as the universe is only 13.8 billion years old, the light would not have had enough time in the whole lifespan of the universe to go from one galaxy to the other. Because nothing can travel faster than light, any exchange between those two places is impossible – and has always been so. Yet the radiation that reaches us proves that the same conditions operate in the galaxy in the eastern sky and in the galaxy in the western sky. This accord is as remarkable as if on his first landfall in the Caribbean, Columbus had encountered Spanish-speaking people in European dress.

In 1980, cosmologists in Moscow and Boston, unaware of each other's work, came up with a theory that solved the problem: the whole of the visible universe was once a neighbourhood. Back then, all the places we see today, billions of light years away from each other in the night sky, were close together. Exchange of information was therefore possible and the same conditions could be established. Then the neighbours were suddenly torn apart. That happened in a phase immediately after the birth of the universe, during which it gained in volume at incredible speed – cosmic inflation.

To call cosmic inflation an explosion of the universe would be a great understatement. An explosion can be photographed with a high-speed camera and has a centre from which everything flies away. As soon as the bang of the explosion is over, the debris blown away in all directions starts to lose speed.

What happened after the Big Bang, however, was on a completely different scale. This inflation had no centre: all of space expanded. The whole event lasted less than a billionth of a billionth of a billionth of a second and in that unimaginably short time, the expansion accelerated so much that the universe immediately grew to at least a billion billion billion times its original size. (Increased by the same factor, the full stop at the end of this sentence would be as big as the distance from Earth to the Andromeda nebula.) Some variants of the theory even envisage a much stronger expansion, by factors such as 10 raised to the power of 10^{10}. (If the full stop at the end of this sentence were enlarged correspondingly, the whole of the universe visible today would not be big enough to accommodate it.) Then the whole thing was over and the comparatively gradual expansion of the cosmos that continues to this day began. So we have inflation to thank for the universe being so large and uniform. It created the prerequisites for our existence.

Before cosmic inflation, the universe was microscopic. The whole of space that we can see today was concentrated in a volume much, much smaller than an atomic nucleus. The whole entity obeyed quantum physics, the laws of nature for the smallest of things, according to which there are chance events that cannot be explained, as Investigator Glock discovered in Chapter 5.

After inflation, the theory says, the visible universe had roughly the diameter of a grapefruit. Compared with its present dimensions that might still seem pretty manageable, but this was a huge leap. After its transformation, the cosmos was a macroscopic world. And because it continued to grow, its constituent parts grew further and further apart. Areas that prior to inflation had been in touch with each other were now out of contact for good. Since then, there has been no common present anymore, but what does remain is the past that connects everything in the visible universe: every point in the night sky was once part of the same microscopic world. That is why conditions are the same throughout the cosmos.

Admittedly this scenario sounds fantastical. And we are still waiting for direct proof of cosmic inflation. We also don't know what precisely set off this inflation. The responsibility is said to lie with the effect of a particle that cosmologists have christened the 'inflaton' – but no one knows what an inflaton is.

However, the theory agrees so beautifully with recent observations that only a few cosmologists still have their doubts about inflation. Traces of the microcosm out of which the universe unfolded still exist today. They appear in the cosmic background radiation, the distribution of which is shown on page 29. This high-resolution photograph that the Planck space observatory recently transmitted to Earth shows minimal fluctuations in the background radiation, a spot pattern in the first light of the world. It shows the typical pattern of a quantum fluctuation – the chance flutter of energy as it inevitably appears in atoms and on even smaller scales. When the universe suddenly expanded, the fluctuation was enlarged to the new, cosmic scale – the way tiny letters printed on a balloon turn into large bold type when someone blows it up. This pattern has impressed itself on the cosmic background radiation. It shines in the sky for all time as an image of the universe before its great transformation.

The pattern of the background radiation was not the only thing the fluctuation, that twitch of energy at the beginning of time, produced. It gave the universe its shape. Through inflation, the fluctuation was magnified to the cosmic scale and turned into a huge blast wave. Thus it determined the way matter was distributed and, as a result, where stars were later located. In the light of the stars we see the signs of another world: the cosmos hiding in the atoms, the tiny something that our universe once was.

If God had a choice to make while creating the world, then He did so at the moment of the inflation. For it was when the microscopic universe, driven by fluctuations, became macroscopic, that it found its shape. That process laid down the characteristics of the universe, which enabled life to arise later on.

The inflation theory, however, contains a remarkable prediction: it allows for the repeated genesis of new forms through different fluctuations that can grow to cosmic dimensions. Each of these forms is a universe in itself. And each is of a different nature. It is even possible that different laws of nature govern the universes that proceed from the various fluctuations. In that case, our cosmos with its order would be only one of many, while elsewhere conditions would be quite different. Thus the inflation theory answers Einstein's question with a counter-question: did God have to make a choice at all? Perhaps He was spared having to make a choice because He could try out everything.

You can picture the universes in their diversity like snowflakes drifting down from a cloud. The snowflakes take on their shape depending on what the atmospheric conditions inside the cloud happen to be. And because the temperature, humidity and wind inside the cloud are constantly and randomly changing, a variety of different snowflakes are formed. In the same way, the individual universes could also differ as they emerge through fluctuations from the microscopic cosmos. Two- or nine-dimensional cosmic spaces could be formed, universes with gravity that crushes everything, or even ones that contain no matter at all. This gathering of universes is called the 'multiverse'. In this multiverse, our own universe would only be a tiny part, a snowflake disappearing in the winter landscape.

Universes are born and many implode again, resembling ice crystals in that as well. The multiverse, on the other hand, is eternal, so it engenders many worlds, most of them hostile to life. But because the multiverse realizes all possibilities, a world such as ours must arise. And only one such universe was suitable to become a place of life – a home to creatures who reflect on its genesis. So it is no wonder, then, that this cosmos appears as if made for us: in all the other universes, there is no one who could delight in it.

We don't know if there really are other universes and whether this prediction of the inflation theory is correct. After all, we can only see a tiny part of our own cosmos. The largest areas of the universe are hidden from us. How then could we find out what lies outside it?

Can there be an outside at all? That depends on what one understands by 'universe'. If that word means everything that exists, then 'outside the universe' is a contradiction in terms. But one can also use 'universe' to designate a cohesive whole of space, time, energy and matter. Perhaps there are other universes alongside ours. Perhaps the Big Bang, to which our universe dates back, was not the beginning of all things. Perhaps our universe was reborn from an earlier universe – we just don't know.

But that doesn't have to be the end of it. Anyone who thinks the latest stage of knowledge is final has always been mistaken. After all, the exploration of the universe has only just begun. It is hardly more than four hundred years since Galileo pointed the first telescope at the skies. Less than a hundred years ago, scientists were still convinced that the universe consisted of the Milky Way alone. Today the estimated number of galaxies in the visible cosmos alone is two billion. And only in the last five decades has the realization that our universe was not always there, that it must have had a beginning, become generally accepted. It would be extremely unlikely that we have by now come to know everything that the universe can tell us about its and our origins.

The traveller in Flammarion's woodcut has broken through the horizon; now the unsuspected beauty of a greater reality has revealed itself to him. He has seen the delights of the Earth, which lies behind him. He has visited its cities, its hills and lakes, and has looked up at the Sun and the stars. But he also knows that at most he has been seeing the surface of things and that everything he has so far viewed could only be one facet of a much more extensive and richer world. He wanted to experience the true nature of the universe.

Humanity is standing at the beginning of the path the traveller has already completed. We have discovered that the universe is much more than the space, time, energy and matter known to us. We know that reality must be quite different from the way it appears to us. We, too, are surrounded by a horizon beyond which lies a more extensive reality. And it is that horizon we are heading towards. As long as we are wise enough to see to it that we continue to exist, we are capable of approaching the mystery and of learning where we come from.

NOTES

Chapter 1

Poe's disparaging opinion of science is astonishing, as he himself looked into the problems of cosmology and came to some original conclusions. In his prose poem 'Eureka', published in 1848, he anticipated ideas that much later became part of science. Poe talked about an expanding universe that began with a 'big bang', and also solved the problem of the dark night sky that is dealt with in Chapter 9 of this book. Did Poe distrust his own enthusiasm for physics?

Genetics, which my fellow speaker at an event at the Munich Literaturhaus in 2000 feared would give us the complete decoding of humanity, has shown us how science poses dozens of new problems for every one it solves. In recent years, for example, the extent to which the living conditions of an organism change its genes has become impressively apparent. The underlying mechanisms are extremely complicated and still largely unknown. And it is out of the question – or as good as – that we will ever succeed in controlling the living conditions of a human being under laboratory conditions and thereby create something like a transparent person with no secrets.

Richard Feynman lived from 1918 to 1988. He was one of the most versatile physicists of the 20th century. He was awarded the Nobel Prize in 1965 for his contributions to quantum field theory, which describes the formation and destruction of elementary particles.

The chronology of the accumulation of oxygen in the atmosphere is described by Lyons, Reinhard and Planawsky in *Nature* 506 (2014). For the perception of light by cyanobacteria, see Schuergers and others, *eLife* 5 (2016).

The metaphor of knowledge as an island in the ocean of ignorance goes back to the distinguished Austro-American physicist Victor Weisskopf (1908–2002). He made his name with contributions to theoretical nuclear physics and campaigned against nuclear weapons testing after the Second World War.

The British biologist J. B. S. (John Burton Sanderson) Haldane lived from 1892 to 1964. He established the field of population genetics, thus providing the modern basis for the theory of evolution. Beyond that, he was a prolific writer, publishing political essays, works of popular science and a children's book. In protest against the attitude of the British government during the 1956 Suez crisis, he emigrated to India, where he spent his final years as head of research at the Indian Statistical Institute in Calcutta.

Chapter 2

The Earth rising over the Moon was photographed by the astronauts of Apollo 11.

The woodcut 'Traveller on the Edge of the World' comes from Camille Flammarion's book *L'atmosphère*, Paris, 1888.

The photograph of the cosmic background radiation comes from the European Planck space probe.

That the cosmic background radiation can indeed be captured with materials from a DIY store was demonstrated by two Berlin schoolboys. Timo Stein and Christopher Förster even managed to measure the properties of background radiation with a satellite dish they had adapted themselves. They described their setup in *Sterne und Weltraum* magazine, July 2008.

The speed at which the universe is expanding can be determined using the speed at which distant galaxies appear to be moving away from Earth. That this speed is greater than the speed of light does not contradict the special theory of relativity. That simply asserts that no object can move faster than light *through space*. In the case of the expansion of the universe, however, it is space itself that is changing.

Chapter 3

Astronomers in the 18th century already knew that light cannot travel infinitely fast. As early as 1729, the English astronomer James Bradley noticed that he had to shift his telescope slightly in order to observe a star that was directly above him – just as a hunter has to aim in front of a running hare for the bullet to hit it. (The hunter would only be able to aim directly at the hare if the bullet were infinitely fast.) Bradley explained this otherwise inexplicable 'aberration of light' with the movement of the Earth around the Sun and the finite speed of light, which he set with astonishing precision at 301,000 kilometres per second. Furthermore, the starlight clearly always arrived at the same speed. Whether the Earth was speeding towards a star or, later in the year, moving away from it, the light did not seem to notice. That was more than surprising.

The wave theory of light could explain many phenomena, such as interference and diffraction, but swiftly ran into trouble because no one could say what the nature of light waves actually was. In order to save their theory, 19th century physicists believed – more out of desperation than with good reason – that a rarefied element filled cosmic space. They called this substance 'the aether'. But in 1881 the American physicist Albert Michelson put an end to the aether. In a famous experiment that he carried out in Potsdam, he sent one ray of light off to the north and a second to the west. If the aether did exist, then the Earth, in its journey around the Sun, would plough through it from east to west, in which case the ray heading westward would feel a kind of headwind and be slowed down in comparison with the one going north. But Michelson established that both rays spread out at the same speed. The aether cannot exist. The

American chemist Edward Morley confirmed this result when he repeated Michelson's experiment in 1887 with an improved configuration.

Robert Friedmann presents an interesting picture of the Nobel Prize Committee that was to honour Einstein in *Europhysics News*, July–August 2005.

The explanation of the special theory of relativity through two flashes of lightning that strike in front of and behind one moving and one stationary observer comes from Einstein himself. He used the image of a train travelling very fast towards one flash of lightning and away from the other. For the passenger in the train, the two flashes appear one after the other, for the observer beside the track simultaneously.

The way time is stretched in a moving object from the point of view of a stationary observer can be made clear by the famous twin paradox: if one of a pair of twin sisters heads off to a distant destination in space as a young girl, she will return as a woman in the prime of life. Her sister who remained on Earth will be an old woman by then. The travelling sister's heartbeat will have slowed down on the way, her days will have grown longer and her first grey hairs appeared later. The reason: both sisters can only measure the same speed of photons. But as seen by the observer on Earth, a flash of light in the spaceship travels a slightly longer distance, because the spaceship is moving. In order to compensate for this difference, time must pass more slowly for the traveller.

Einstein made his remark about not being sure whether

'the Lord God was pulling his leg or not' when he discovered the equivalence of mass and energy to the Swiss physics teacher Conrad Habicht. The two men had got to know each other in Schaffhausen, where in 1901 Einstein was temporarily earning his living as a private tutor. Habicht and Einstein remained friends throughout their lives.

Chapter 4

With his work, Pierre-Simon Laplace (1749–1827) created the basis for probability theory and analysis. He also proved that planetary orbits are in general stable, which was doubted at the time. He also served as Napoleon's minister of the interior, an office in which he displayed such unbearable pedantry that Napoleon dismissed him after only six weeks.

For those interested, the Schrödinger equation is given below:

$$i\hbar \frac{\partial}{\partial t} \Psi(r,t) = \hat{H}\Psi(r,t)$$

The equation indicates how the N-particle wave function $\Psi(r,t)$ develops in time. This wave function, as a solution of the Schrödinger equation, fully describes the quantum mechanical state of the system. It allows us, for example, to deduce the probability of the location of each particle.

i is the imaginary unit, $\hbar$ the reduced quantum of action, r the position vector of the N particles $r = (r_1, r_2, \ldots r_N)$ and, as usual, t stands for time and $\frac{\partial}{\partial t}$ for the partial derivative with respect to time. $\hat{H}$ is the Hamiltonian for the system of $\mathcal{N}$ particles that consists of $\mathcal{N}_e$ electrons, and $\mathcal{N}_A$ atomic nuclei:

$$\hat{H} = -\sum_{i=1}^{N_i} \frac{\hbar}{2m} \nabla_i^2 - \sum_{\alpha=1}^{N_A} \frac{\hbar}{2M_\alpha} \nabla_\alpha^2 - \sum_{i=1}^{N_e} \sum_{\alpha=1}^{N_A} \frac{Z_\alpha e^2}{|r_i - r_\alpha|} + \sum_{i=1}^{N_e} \sum_{j=1}^{i} \frac{e^2}{|r_i - r_j|} + \sum_{\alpha=1}^{N_A} \sum_{\beta=1}^{\alpha} \frac{Z_\alpha Z_\beta e^2}{|r_\alpha - r_\beta|}$$

m is the mass of the electron and e the electron charge. M_a is the mass and Z_a the number of protons in the nucleus a. ∇_i stands for the derivative with respect to the coordinates of location of particle i. The first sum in the Hamiltonian describes the kinetic energy of the electrons, the second sum the kinetic energy of the nuclei, the third sum the Coulomb interaction between the electrons and nuclei, the fourth the Coulomb interaction between the electrons, the fifth the Coulomb interaction between the nuclei. Relativistic effects have been ignored because they have no part to play under everyday conditions. In order to make the whole clearer, non-relativistic gravity and the effect of electromagnetic radiation have been omitted, but there is no problem adding them. With its Schrödinger equation and the wave function at a random point in time t_0 as the initial condition, the dynamics of the system are completely determined.

We have Claude Shannon (1916–2001) to thank for modern information theory. He was a scientist with an unusually wide range of interests: when he was 21, he developed the fundamentals of the integrated circuit on which today's computer technology is based; during the Second World War, he worked on cryptography; and in 1950 he built an electronic mouse that could find its way around a maze and is regarded as the first robot with a capacity to learn. In his free time, Shannon amused himself by riding a unicycle, juggling and using his understanding of game theory to win large sums of money at blackjack in Las Vegas. He published his groundbreaking thoughts on the possibilities of chess computers in *The London, Edinburgh and Dublin Philosophical Magazine and Journal of Science* 41, 1950.

For an analysis of the mistakes in the forecast of the Lorenz hurricane see: Majewski and others, *Monthly Weather Review* 130, 2002.

Chapter 5

In the literature on quantum mechanics, the two observers of entangled states who are apart from each other are generally called Alice and Bob, like Mr and Mrs Aspect in the crime story.

Alain Aspect, born in 1947, worked in the optics laboratory of the Université Paris Sud. The experiments in which entangled photons were sent from one island in the Canaries to another, or to Austria and China, go back to the Viennese physicist Anton Zeilinger, born in 1945. He too was a pioneer in quantum optics. Aspect and Zeilinger received all the important awards for physics, apart from the Nobel Prize.

Entanglement occurs everywhere in nature; the phenomenon always appears whenever elementary particles interact randomly. For the experiments described, the entanglement has to be established in a controlled way. The simplest way to do this is by sending a laser beam into special crystals, such as barium borate. When this is done, the atoms in the crystal become excited so that the original ray of light is divided into two partial rays. Now every photon in one partial ray has an entangled partner in the other.

Spin is a characteristic of elementary particles that does not exist in classical physics. The spin of electrons and other particles with mass is measured by an experimental arrangement that was designed in Frankfurt by the German physicists Otto Stern and Walter Gerlach in 1922. The particle is sent through a magnetic field; because spin links up with the magnetic field, the particle is then diverted either up or down. The spin of a photon can be established with polarisation filters.

In Aspect's experiment, the spins of two paired photons were opposed to each other. In order to get corresponding answers, an upward spin is measured in one photon, a downward spin in the other, or vice versa.

One might suspect that entangled particles allow effects to be transferred faster than light and that entanglement therefore contradicts the special theory of relativity, according to which no effect can spread faster than light. But there is no contradiction. Cause-and-effect means that an earlier, known event sets off a later event. In quantum mechanics, however, there are no events that are known in advance. Events occur randomly, and only when a system is subjected to a measurement at that. Therefore two observers

– let's call them Alice and Bob – cannot send a message to each other by entangled particles alone. When Alice establishes by measuring her particle that there has been an event, all she knows is that Bob must notice a corresponding event on his particle. But she is unable to manipulate her particle in such a way that she can say in advance what event Bob will observe. Nor does she know what will happen on her side. So because Alice does not know what events will occur, she cannot cause an effect on Bob's side. That means, therefore, that the simple fact of entanglement is not enough to transmit information; the special theory of relativity remains in force.

John Bell, born 1928, worked on particle physics at the CERN European research centre in Geneva. He died from a stroke in 1990. That same year, he was nominated for the Nobel Prize.

The three questions in the crime story correspond to three different measurements of the spin of two entangled particles, where the axes of the three measurements are each at an angle of 120 degrees to one another. In this case, the result brought about by Glock corresponds exactly to the argument deduced by Bell: on the assumption of a hidden plan, the measurement will agree with a probability of 5/9, but when such a plan is absent with a probability of 1/2.

The idea that space develops through quantum mechanical entanglement was developed within the framework of string theory. Much discussed in recent years, string theory is an attempt to unite quantum mechanics and the theory of relativity. High-profile supporters of the idea that spacetime arose through entanglement are the American string theorists Juan Maldacena and Leonard Susskind; see for example Maldacena and Susskind in *Fortschritte der Physik* 61 (2013).

Chapter 6

It was contemporary philosopher Nick Bostrom at Oxford University who proposed that the reality we experience is a computer simulation from a post-human civilisation.

Buddhist conceptions of reality were already described in the discourses of the Buddha and later developed further. Central to them is a concept called *shunyata*; this word can best be translated from the Sanskrit as 'emptiness'. See Nishitani Keiji, *Religion and Nothingness*, University of California Press, Berkeley, 1982.

In classical western philosophy, the Roman poet Lucretius speaks of 'simulacra', in line with the epistemology of the 1st century BC. These simulacra are very fine skins that peel off the outer surface of all things, leaving an impression on the eye. That is how we come by our perceptions, Lucretius explains, in his didactic poem 'De rerum natura' (On the Nature of Things).

Filmmakers of the 20th century have taken up this concept. In Rainer Werner Fassbinder's *World on a Wire* (1973), which in many respects anticipated *The Matrix*, the characters are forced to accept that the world is just a 'simulacrum' – an illusion created by a computer. Neo, the main character of *The Matrix*, hides his secret software in a book by the French philosopher Baudrillard with the title *Simulacra and Simulation*.

Ernest Rutherford's dark premonition that with an atomic chain reaction one could 'make the world go up in smoke' is reported by Richard Rhodes in his fascinating chronicle *The Making of the Atom Bomb*, 1986.

As with all experimental data, there is a detection limit for the measurements of particles. At the moment it is at 10^{-18} m. Given that limit, we can rule out elementary particles having dimension. In current theories, elementary particles are assumed to be points.

The Higgs field is a so-called scalar field. Scalar fields attribute a unique number to every point in a space. So the distribution of temperature in a room, for example, is indicated by a scalar field. All known particles apart from the Higgs, on the other hand, have to be described by vector fields – they have linear momentum and an intrinsic angular momentum, the spin. The Higgs lacks those characteristics, which means it is a fundamentally new type of particle. Whether there is just one kind or several kinds of Higgs particles remains an open question.

Leibniz speculated about the void in *The Principles of Nature and Grace Based on Reason* (1714).

Chapter 7

The gravity lens Abell 1689 is formed by the direct and the Einstein Cross by the indirect effect of dark matter. With Abell 1689, it is the dark matter itself that diverts the light from the surrounding stars. The galaxy at the centre of the Einstein Cross warps the light of the quasar behind it above all through the gravitational effect of the stars concentrated in the centre of the galaxy bthe stars concentrated in the centre of the galaxy. But this tight packing of stars is only stable because it is held together by a powerful halo of dark matter – see Trott and others, *Monthly Notes of the Royal Astronomical Society* 401 (2010).

The photograph of Abell 1689 on page 134 was transmitted to Earth from the Hubble space telescope.

Dokkum and others have described the dark twin galaxy of the Milky Way in *Astrophysical Journal Letters* 826 (2006).

An estimate of the amount of dark matter constantly hitting and passing through the Earth is given by Gaitskell and others in *Annual Review of Nuclear and Particle Science* 54 (2004).

Lisa Randall and Matthew Reece describe their theory in *Physical Review Letters* 112 (2014). See also Lisa Randall's book *Dark Matter and the Dinosaurs*, HarperCollins, 2015. The clustering of cometary impacts asserted by Randall and others is disputed. For example, in 2011 scientists at the Heidelberg Max Planck Institute for Astronomy came to the conclusion that the apparent periodicity of the impacts is just the result of statistical artefacts and that there is no such correlation.

Muons, which once confused Rabi and his colleagues, are exotic particles with a life of only a few millionths of a second. Under normal circumstances, muons are unimportant for the behaviour of matter; the fact that they exist disrupted the much simpler construction set of particle physics that was known at the time. They also have their place in the Standard Model.

Einstein believed in a static universe, but recognized that his field equations for the general theory of relativity generally resulted in an expanding universe. He therefore added a constant to his equation that was intended to counteract the expansion. This constant describes a pressure of empty space and corresponds to what we today call dark energy. However, Einstein quickly realized that the cosmological constant could not do what he hoped it would. The statistical solutions of the modified equations are not stable against perturbations. A minor change in the parameters immediately results again in an expanding universe. Therefore he withdrew from his attempt. Today we know that his presumption of a stable universe was wrong.

The dark energy in the universe corresponds to the energy of seven protons per cubic metre. The dark energy of an empty space the size of all the oceans is thus far less than the energy of a drop of water.

Chapter 8

The only known exception to the indifference of the fundamental laws of nature to the direction of time concerns the decay of certain building blocks of the atomic nucleus under the influence of the weak interaction. With these processes of decay, the validity of the laws of nature during time reversal is only retained when at the same time space is mirrored and matter replaced by antimatter. This so-called CPT invariance is a fascinating phenomenon, but otherwise of no importance. As long as there is no degradation of the nucleus, the laws of nature allow time reversal.

Boltzmann's most vehement opponent was the Viennese physicist Ernst Mach (1838–1916), whose work on mechanics prepared the way for the theory of relativity. The comparison with a bullfight was made by the Munich physicist Arnold Sommerfeld (1868–1951), a pioneer of atomic physics.

The cellular mechanism of hair turning grey is described by Emi Nashimura and others in *Science* 307 (2005).

The second law of thermodynamics can be formulated in various ways. The formulation of the second law given here is especially vivid and comes from the renowned German physicist Max Planck (1858–1947). Strictly speaking, it ought to say 'almost never'. Fluctuations in the system can in fact reduce its entropy temporarily, but in the long term they are irrelevant.

Chapter 9

The observation that a night sky that is infinite in time and space cannot be dark is called Olbers' paradox after Heinrich Olbers (1758–1840), a doctor and astronomer from Bremen, who described the problem in 1823. However, the issue was already known to Kepler. Olbers' formulation rests on the plausible assumption that the shining objects in the universe are, on the whole, regularly distributed. The paradox cannot be resolved with the idea that cosmic clouds of gas or dust absorb the radiation. As the British astronomer John Herschel (1792–1871) recognized, after a certain time a thermal balance must ensue in which the cloud emits as much radiation energy as it absorbs.

Edwin Hubble (1889–1953) spent his whole working life at the Mount Wilson Observatory in the mountains near Los Angeles. When he took up his position there in 1919, astronomers still believed that the universe consisted of the Milky Way alone. Hubble was the first to realize that there are galaxies outside our own. In 1922 and 1923, he made the seminal discovery that objects such as the Andromeda nebula were much too far away to be part of the Milky Way. Six years later, he detected the red shift in the light of distant heavenly bodies, from which he concluded that the universe was expanding. In his honour, the correlation between the distance to a galaxy and its recessional velocity is today called Hubble's Law.

Incidentally, Hubble explained his measurements

with the well-known Doppler effect: he claimed that the wavelength of light coming from distant and therefore fast-moving stars changed in the same way that the siren of an ambulance will sound slightly deeper if is moving away than if it is approaching. However, the analogy doesn't work: the ambulance is moving through space, while the red shift of the starlight comes from the fact that space itself is changing its size. Hubble had misunderstood the theory of relativity.

Even though the whole of modern cosmology rests on his pioneering achievements, Hubble was refused the Nobel Prize because at the time the prize committee was not yet allowed to honour work in astronomy.

The warp of the universe can be worked out from data on the cosmic background radiation collected by the WMAP and Planck space probes. To do so, physicists also exploit the fact that the cosmic background radiation shows tiny irregularities. These so-called fluctuations came from the pressure waves that were running across the universe when the cosmic background radiation arose. The measurements of such waves can be derived from the acoustic velocity in the early universe, which comes from the known natural constants. The figures calculated for the fluctuations are then compared with the data sent by the space probes. In a warped universe, the fluctuations would appear as if distorted through a lens, that is to say, they would look larger or smaller than expected. In fact, however, both the calculated and the observed measurements of the fluctuations agree precisely. From that it follows that the universe is flat.

Various estimates of the minimum size of the universe are given by Vardanya, Trotta and Silk in *Monthly Notices of the Royal Astronomical Society: Letters* 413 (2011), Caspar Douspi and Ferreira in *Physical Review D* 68 (2003), as well as Silk in his book *The Infinite Cosmos: Questions from the Frontiers of Cosmology*, OUP, London & New York (2006).

The reflections on probability during the frequent repetition of an experiment involving chance are easy to reproduce mathematically using a game of dice as an example. Let p be the probability for a single event X. X can, for example, stand for the six being thrown; with a standard six-sided die, $p = 1/6$. Correspondingly, the probability of another number being thrown is $q = 1 - p = 5/6$. The probability of no six turning up after n throws is q^n; the probability of a six being thrown at least once is $1 - q^n$. As n increases, the latter probability approaches 1. For two sixes being thrown, $p = 1/36$, the further calculation being analogous. On the basis of the exponential correlation, the single probability p is ultimately of no significance. As long as the frequency of the repetitions n is big enough, the probability of X occurring at least once for an arbitrarily small $p > 0$ always comes arbitrarily close to 1.

Giordano Bruno's reaction to the verdict is documented in the collection of sources for the proceedings of the Inquisition against him edited by Angelo Mercati (1942).

In the context of physics today, the Russian-American cosmologist Alexander Vilenkin in particular has speculated on the inevitability of events and repetitions in an infinitely large universe. He sees his reflections in the context of present-day inflation cosmology, as described in the following chapter. See Garriga and Vilenkin, *Physical Review D* 64 (2001).

Chapter 10

What speaks for the idea of life starting in tidal pools is that the earliest known fossil remains of single-celled organisms are to be found in sediments that once built up on beaches and in the shallow waters of the oceans. The regular alternation between inundation and evaporation meant that organic substances built up in such areas. A rival hypothesis assumes that life began around volcanic vents on the ocean floor. However, that theory cannot explain how the self-replicating molecules could concentrate in cells.

The improbable conditions that had to arise for life to develop on Earth have been described in detail by Ward and Brownlee in their book *Rare Earth* (2005). How a protoplanet could approach the young Earth in such a way that they collided and the Moon developed out of the mass of stone debris has been computed by the mathematician Edward Belbruno and the astrophysicist Richard Gott III. The work of these American scientists, published in *The Astronomical Journal* 129 (2016), also contains a summary of current knowledge about the genesis of the Moon.

Frank and Sullivan demonstrated their assessment of the chance of life on Earth in *Astrobiology* 16 (2016). The two authors explicitly investigated the probability of a civilisation with highly developed technology existing in the visible universe, because the radio signals emitted by such a civilisation could possibly be received. However, the argument given by Frank and Sullivan could just as well be used to assess the probability of the development of any kind of life. For the frequency of stars similar to the Sun around which planets are orbiting in the life-friendly zone, see Petigura and Marcy in *Proceedings of the National Academy of Sciences* 110 (2013).

For that matter, there is no reason to think that our solar system is among the more life-friendly ones in the universe. Only one planet, Earth, lies clearly within the habitable zone, where liquid water and a temperate climate make life as we know it possible. By contrast, 40 light years away from us the planetary system TRAPPIST-1, the discovery of which was announced in February 2017, contains at least three habitable planets that would allow life. When precursors of life emerge on a habitable planet, these biomolecules are almost certain to spread to the neighbouring habitable planets via meteorites. Therefore, the probability of the emergence of life multiplies in a planetary system of this type. See Manavasi Lingam and Abraham Loeb in *Proceedings of the National Academy of Sciences* (2017).

The fact that various areas of the universe are not in contact with each other is called the horizon problem. One might believe that the horizon problem would be solved by the expansion of the cosmos, because distances between places in the early universe were smaller, so larger areas of the universe could influence each other. But in fact, expansion exacerbates the horizon problem, because space was expanding faster in the past than it is now. In an earlier age – seven billion years after the Big Bang, for example – when the cosmos was smaller, the two galaxies were indeed closer together, but light also had less time to get from here to there. In our example, there were seven billion years available for the journey back then, a little more than half as much time as today. But the distance between the two galaxies then was considerably greater than half of what it is now, because the closer we get to the Big Bang, the faster the universe was expanding. The further in the past, the faster the two galaxies were moving away from each other. If exchange between them is impossible today, it was all the more impossible back then.

Like all theories of modern fundamental physics, the inflation theory is a field theory. The dynamics of inflation are attributed to the effect of a spatially distributed factor. This variable of an unknown nature is the inflation field. Only very general assumptions can be made about it, which means there are a number of rival inflation theories. The inflaton is allocated to the inflation field and corresponds to a quantized excitation of the inflation field. The common inflation theories describe the inflation field as a scalar field, similar to the Higgs field discovered in 2012 at the CERN accelerator that imparts mass to other elementary particles. It is suspected that there is a close connection between these two mechanisms.

That certain parameters of inflation can be read from the radiation of the stars was shown by the Sloan Digital Sky Survey. In the course of the most comprehensive survey of the night sky so far, since 1998 the positions and spectra of over 200,000 galaxies have been measured with an automated telescope. The inflation theory offers the best explanation for the recorded data. For this, see Tegmark and others, *Physical Review D* 69 (2004).

An estimate of the number of galaxies in the visible cosmos is given by Conselice and others in *Astrophysical Journal* 830 (2016).

Thanks

For conversations, criticism and support during the development of this book, I would like to thank Franz-Stefan Bauer, Béa Beste, Ulrike Bartholomäus, Volker Foertsch, Alfio Furnari, Gabriele Hoffmann, Ben Moore, Viatcheslav Mukhanov, Peter Knippertz, Matthias Landwehr, Thomas de Padova, Martin Rees, Alexandra Rigos, Efstratios Rigos, Wolfgang Schneider, Nina Sillem, Herbert Wagner and Steven Weinberg.

About the author

Stefan Klein (b.1965) is Germany's bestselling science author. He studied physics and analytical philosophy in Munich, Grenoble and Freiburg, and conducted research in the field of theoretical biophysics. He turned to writing because he 'wanted to inspire people with a reality that is more exciting than any crime novel'. His book *The Science of Happiness* (2002) topped all the German bestseller lists. His most recent bestseller, *Träume* ('Dreams'), won the prestigious Deutsche Lesepreis 2016.